Thomas Walther

That Bond that matters

Bibliographic information of the German National Library

The German National Library has listed this publication in the German National Bibliography. Detailed bibliographic data are available from dnb.dnb.de.

Copyright

Acknowledgements

The author wishes to express sincere gratitude to the many scientific authors and researchers whose work is cited in this book. Their scholarship, careful analysis and valuable contributions have provided an important foundation for this publication. The author has learned a great deal from their work and is deeply appreciative of the knowledge they have shared.

Spelling, grammar, and final proofreading were supported using the Grammarly proofreader.

Contact: technology-research@posteo.eu
Publisher: BoD · Books on Demand GmbH, Überseering 33, 22297 Hamburg, bod@bod.de

Print: Libri Plureos GmbH, Friedensallee 273, 22763 Hamburg

ISBN: 978-3-6963-7802-8

Preface

Corrugated board has captivated me professionally for more than thirty years.

My career began with trials in direct offset printing on corrugated board, seeking marketable print quality. Persistent striping and crushing revealed intricacies beyond textbook knowledge, even though the project fell short.

Corrugated board has remained central to my work ever since. Over the years, I have approached it from many perspectives: as a printing substrate, a bonding substrate, a folding material, and a transport packaging material. Each time I believed I had mastered it, new complexities emerged. I continue to learn from it today; this ongoing learning is what keeps the material so engaging.

Against this backdrop of continuous discovery, I offer this book. It does not claim to have all the answers. Instead, it is an attempt to make visible the connections that tend to get lost in day-to-day practice: how micrometres at the corrugator become millimetre deviations at the packaging line; why a glue line fails even though the adhesive is perfectly sound; and why reducing consumption sometimes improves quality rather than compromising it.

I wrote this book originally for myself—to clarify my understanding. Chapters evolved from private notes. The intention to share came later, as I wanted others working with the material to benefit.

This book is written with three main audiences in mind: the converter production manager responsible for line results, the packaging engineer who specifies materials, and the adhesive and coating specialist focused on technical detail. If you work in an adjacent field, such as paper mills, OEMs, or retail, you will find context tailored for you in the Introduction and Glossary, ensuring relevance across the value chain.

A note on transparency: my professional home is Baumer hhs, a supplier of adhesive application and monitoring technology. Specific systems from that company appear in later chapters as illustrative examples. I have taken care to describe the underlying physics and engineering principles without framing the discussion around product specifications; wherever a named product is referenced, the reader can substitute an equivalent system from another supplier without losing the argument.

In that same spirit, the book is sold at cost. The price is set to cover the expenses incurred in producing it, and no more. I deliberately chose the route of a book-on-demand publisher rather than a conventional publishing house because it keeps the cost to the reader low and aligns with the spirit in which the book was written: as a contribution to a shared field of practice, not as a commercial product.

If you see certain things differently, that is a good thing. If you would like to add something or share a comment, I welcome your feedback. A technical book thrives on dialogue, and for me, that dialogue has only just begun with this book, not ended.

My thanks go to the colleagues at Baumer hhs, whose daily work in adhesive application and quality monitoring made the practical grounding of this book possible. I am equally grateful to the many corrugated board converters who have given me insight into their processes over the years and have helped shape my understanding of this material.

Thomas Walther
Krefeld, Spring 2026

Introduction

Corrugated board is one of the most sustainable packaging materials in the world, with recycling rates exceeding 90 per cent in Europe and a closed material cycle that is unmatched in the packaging industry. It is also one of the most underestimated. The industry measures grammage, bursting strength and edge crush values, yet the real causes of complaints, machine downtime and material waste lie deeper: in the interaction between surface quality, crease accuracy and glue line integrity.

When these relationships are not properly understood and controlled, the consequences are far-reaching. Converting lines come to a standstill because warped blanks can no longer be erected. Glue lines fail under the spring-back force of the material before the adhesive has even set. Incorrectly dimensioned packaging damages the product or fails retail acceptance testing. At the end of the chain come complaints, penalty charges and the loss of customer confidence, costs that can amount to many times the value of the material itself.

This book follows a corrugated board blank from its formation at the corrugator, through converting at the folder-gluer, to the finished pack on the pallet. It shows how washboarding at the micrometre scale disrupts adhesive distribution, how warp places the glue line under peel stress, and why a wandering crease line results, further down the chain, in a box that can no longer be erected. But it also shows how this cascade of failures can be broken with the right means: through stitching that reinforces the glue line rather than weakening it, and through waterscoring that removes the spring-back force from the material before it can cause damage.

The relationships presented here are grounded in scientific research, practical experience and measurement-based evidence. They are addressed to everyone who manufactures, converts or specifies corrugated board and wants to understand what truly holds this material together.

Each chapter begins with an introduction card that provides a brief overview and orientation for the reader; for further guidance, see the section **Structure: “How to Use the Introduction Cards”**.

Scope

This book focuses on box production at the flexo folder-gluer and on the behaviour of the finished pack on packing lines, on the pallet and in the supply chain. It addresses the corrugator only to the extent that its process outputs, washboarding, warp, paper moisture and bond quality, influence subsequent converting.

It is not a handbook of corrugated board manufacture, but an analysis of the value chain from folder-gluer bonding to the retail shelf.

Each chapter draws on published research, patents and field evidence. Where the evidence base is thin, this is stated openly. The analogies used throughout, from aerospace engineering and from medicine, are there to aid understanding, not to oversimplify. The underlying mechanisms are described in their full depth.

Structure

The book follows a medical logic: from anatomy and pathology through symptom and diagnosis to treatment and prognosis.

Part I - Anatomy (Chapters 1 and 2). Chapter 1 establishes the stakes: why millimetre deviations at the manufacturer's joint can cascade through the entire supply chain. Chapter 2 traces the origin of defects at the corrugator: washboarding, warp, striping, crushing and their interactions with moisture.

Part II - The Consequences, or Pathology (Chapters 3 to 5). Chapter 3 follows the defects to the folder-gluer, where they converge at the manufacturer's joint. Chapter 4 quantifies the economic costs of dimensional deviation, from case erector stoppages to pallet rejection and retail chargebacks. Chapter 5 examines the ecological dimension: why the environmental leverage of packaging lies in its protective function, not in its material weight.

Part III - Root Cause Analysis, or Diagnosis (Chapters 6 to 9). Chapter 6 provides a systematic diagnosis: the causal chain, the interaction matrix, and the insight that variation, not systematic deviation, is the true enemy. Chapter 7 reads the causes of folding variation across the converting line: the lateral offset at the infeed, the skewed feed, the cut position relative to the flute, the moisture gradient across the pallet and the two kinds of crushing. Chapter 8 turns to the secondary fracture next to the crease, the so-called Nebenbruch, and maps its six causes from tool overdrive through flute phase to false scoring. Chapter 9 addresses adhesive application defects, their patterns, their causes, and a three-level safety concept for prevention, detection and response.

Part IV - The Treatment, or Therapy (Chapters 10 to 12). Chapter 10 introduces stitching: interrupted adhesive application as a damage-tolerance strategy for the manufacturer's joint, drawing on the principle of crack arrest from aerospace engineering. Chapter 11 presents waterscoring: targeted crease-line conditioning that reduces spring-back forces, homogenises crease quality and prevents liner fracture. Chapter 12 extends the waterscoring concept beyond the folder-gluer to packing-line applications.

Part V - Synthesis and Outlook (Chapter 13). The final chapter brings the threads together. It describes the synergy between stitching and waterscoring, defines the limits of both interventions, and provides a staged implementation sequence with an economic assessment. It closes with an integrated response to the challenge posed by material reduction regulation. The central argument of this book can be stated in a single sentence:

> **The greatest threat to corrugated packaging is not the individual defect, but uncontrolled variation.**
>
> **The greatest opportunity lies not in more forceful intervention, but in more intelligent conditioning.**

The chapters that follow provide the evidence for this argument.

How to Use the Introduction Cards

The introduction cards at the beginning of each chapter are intended as a reading guide, not as a substitute for the chapter itself. They summarise the central question, the practical relevance, and the main economic, technical, ecological or regulatory implications. Their purpose is to help readers decide how deeply they need to enter a chapter according to their role, whether as production manager, line operator, quality manager, packaging engineer, or specialist supplier. Readers who need a quick orientation may use the cards to follow the argument of the book. Readers facing a specific problem should use them to identify where the full technical discussion begins.

Content

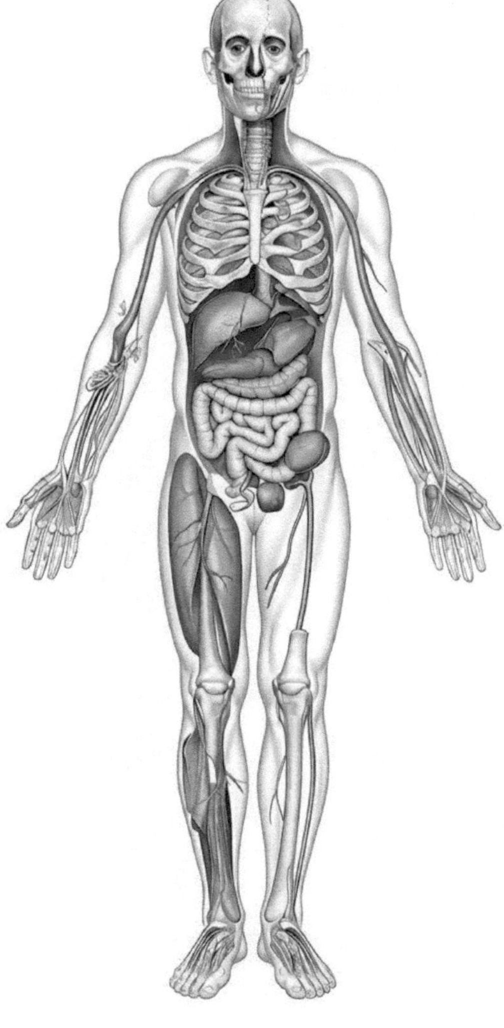

Part I – The Anatomy

Chapter 1: Why One Millimetre Can Break a Supply Chain

Introduction card

A one-millimetre dimensional deviation can stop a case erector, disturb pallet stability, and create costs far beyond the value of the box itself. This chapter explains why small geometric errors matter, where their effects first become visible, and how they move from the folder-gluer to the retail shelf.

"Why does a one-millimetre dimensional deviation at the manufacturer's joint cascade across four nodes of the supply chain, and why is the case erector often the first place where the cost compounds?"

Economic impact: Dimensional deviations create costs at four main stations. At the case erector, small errors cause pop-opens and jams, leading to micro-stops of 30 to 180 seconds. These can add up to 15 to 30 minutes per shift. When the case erector stops, the filling and labelling lines often stop as well. At pallet level, boxes outside specification stack unevenly, concentrate loads at the corners, and can reduce real-world BCT by 30 to 50 per cent compared with laboratory values. At retail, the same defects can lead to returns, chargebacks, and supplier-rating penalties that exceed the direct cost of the defective pack.

Quality: Each defect has a measurable parameter or a diagnostic signature. Washboarding is measured as surface amplitude and should remain below

50 microns; above 100 microns, it becomes process-critical. Warp appears as millimetre-scale board curvature; the target moisture differential between outer and inner liner is below 1 per cent. Striping appears as periodic ink-density variation in post-print and as flute-height variation across the web, which should stay within ±2 per cent of nominal. Crushing requires special attention because the board can appear acceptable while its load-bearing capacity has already been reduced, the only one of the four defects without a reliable visible signature.

Process: Line performance is governed by variation, not by averages alone. The converter's task is to understand and manage variation, not simply to remove it. This principle guides the later interventions. Stitching manages variation by design. Waterscoring narrows the range of variation where it matters most. This chapter establishes the framework for the rest of the book.

Limitations: This chapter defines the problem. It does not yet provide the solution. Solutions are introduced in Part IV, especially Chapters 10 to 12. The focus here is on diagnostic terms, the cost cascade, and the route through the later chapters.

Read in full: every reader of the book.

Prerequisite: none.

Follow-up: Chapter 2, where the defects begin.

On the Hidden Vulnerability of Corrugated Packaging

Imagine a filling line that erects, fills, closes, and palletises 200 packs per hour. At this pace, every stoppage is noticeable, yet small, repeated disruptions – like a misaligned flat or missed suction – halt the erector and back up the filling machine, eroding overall efficiency.

Products queue. A line operator removes the defective blank and restarts the machine. These events often go unlogged, so production reports look normal while the weekly OEE figure reveals hidden losses.

In the modern packaging industry, this critical connection is often underestimated. Most attention is drawn to major topics: machine speed, material costs, and sustainability. Less visible, yet more decisive, is an inconspicuous detail—dimensional accuracy. The real question? Whether a corrugated box, after creasing, folding, and glueing on the flexo folder-gluer (FFG), matches its intended dimensions.

This chapter shows what can happen when dimensional accuracy is missing. Millimetre deviations can cost tens of thousands of euros. The text traces hidden risks from the folder-gluer through the packing line, pallet and loading bay. It introduces two main principles for the book: damage tolerance and targeted conditioning.

1.1 The Packaging as an Interdependent Component

A corrugated box is not a standalone product. It is a component within a system of machines, processes and physical boundary conditions. At the flexo folder-gluer (FFG, a machine that creases, prints, folds, and glues box blanks), the flat blank (unassembled box) is creased, printed, folded and bonded at the manufacturer's joint (MJ, the seam where the box is glued). The result is a knocked-down flat (KDF, a flat-packed box) that is only erected into a three-dimensional box at the end customer's site or on a packing line.

That erection is the first moment of truth. The box must unfold with precision in a fraction of a second, be picked up by suction cups or grippers, fed into the forming station, and assume exactly the specified internal dimensions there. The tolerances (acceptable dimensional variation) in modern case erectors (machines that automatically form boxes) are typically 3 to 5 millimetres for standard RSC boxes (Regular Slotted Container, the most common box design)[86].

Though minor at first glance, tolerance stacking—the accumulation of tiny deviations—causes serious trouble. Picture this: a two-millimetre KDF width deviation. Add two more from fishtailing. Then throw in uneven creasing on top. Together, they overwhelm the erector.

This is called "starve and block" in the packaging machinery industry. One machine is blocked, while another is starved of input. The erector jams (blocks). Upstream, the filling equipment is no longer supplied with boxes (starves). On fast packing lines, such thirty-second stoppages typically go unlogged. In the author's operational experience, they occur about ten to twenty times per shift [84], a figure consistent with stoppage ranges published by case-erector suppliers. The precise frequency varies significantly with line speed, substrate (the material used), and format. Together, these stoppages add up to hours of hidden OEE (Overall Equipment Effectiveness) losses each week.

1.2 Anatomy of a Failure: Fishtailing and Off-Dimension Boxes

The finished box's surface appearance takes two dominant forms.

The first is the off-dimension box: although the panels meet, the gap at the manufacturer's joint (glue seam) is either too wide or too narrow along its whole length.

The second is the fishtail: the two edges of the joint are not parallel, but splay outward like a fish's tail, so that one end of the joint shows a wider gap than the other.

Both produce the same downstream symptom – a box that no longer engages with the case erector – but they arise from different mechanisms. A deep analysis of these two signatures, their upstream causes and the diagnostic patterns they leave in the bundle follows in Chapter 7, where the causes of folding variation are fully treated. Chapter 8 then turns to the secondary fracture that sometimes forms next to the primary crease and feeds into the same fold deviation.

Costs start before anyone understands the mechanism. Each off-dimension or fishtailed box from the folder-gluer triggers a chain of downstream costs. These costs are explored in this chapter and in Chapter 4.

1.3 The Multiplier Effect: From the Production Line to the Loading Bay

Dimensional deviations are not local problems. They multiply costs at every stage of the value chain. To see why, follow a defective box's journey.

At the folder-gluer

The defect starts here: creasing is uneven, and spring-back forces vary. As a result, the fold misaligns from the crease line. When the manufacturer's joint is bonded under tension, the glue line is under peel stress from the start. In the delivery stacker, some boxes spring open because the glue line cannot resist the spring-back force.

At the packing line

Case erectors and filling stations are designed for boxes with accurate dimensions. Off-dimension boxes jam, tip, or fail to fully close. When the erector halts, the entire downstream line is affected. Upstream equipment backs up with each disruption.

On the pallet

The laboratory benchmark for stacking is the Box Compression Test (BCT), which measures the maximum weight a box can withstand before collapse.

Flat platens are used, the load is centred, and humidity is controlled [40]. Yet, reality is different.

> Sidebar: The Box Compression Test in two sentences
>
> The Box Compression Test (BCT) measures the vertical load a corrugated box can withstand before its walls buckle; it is the standard laboratory test of stacking strength. Because lab conditions are ideal—with flat platens, a centred load, and controlled humidity—the BCT value serves as an upper bound. However, the load a box encounters in the supply chain is almost always lower.

At the retailer's loading bay

The moment the pallet leaves the lorry marks the second moment of truth. Even if less than one per cent of the goods are affected, major retail chains reject entire pallets if a single box shows damage. Unloading windows are often measured in minutes, which makes inspection impractical. The result: anything that appears damaged is rejected in its entirety.

Major retailers' chargeback programmes turn packaging failures into a direct financial liability for suppliers[36]. Administrative charges, per-unit surcharges, and return freight costs can quickly add up to thousands of euros per rejected pallet.

1.4 The Hidden Environmental Costs

The economic costs are severe, but they represent only the visible tip of the iceberg. Beneath them lies an environmental cascade that still receives too little attention in industry debate.

When a corrugated package fails, the damage does not stay with the box; instead, products nearby, including the contents, are damaged or contaminated. Even a single dented carton can change load distribution throughout the stack. In many cases, an entire pallet is rejected, even though most of its contents are in good condition.

In the food sector, this is especially true. For most food categories, packaging manufacturing accounts for only a small share of the product's total climate impact—typically about 5 per cent, though it varies widely [102]. When packaging fails, and contents are discarded, every upstream resource is wasted: agricultural land, water, energy, transport, and cold-chain capacity. Packaging's environmental leverage lies not in its material weight, but in its protective function.

There is another overlooked effect: soiled corrugated board disrupts recycling. Corrugated board is recycled at over 90 per cent in Europe, one of the highest rates among packaging materials [50]. Yet a single bale with food residues can ruin an entire batch. Moisture in recycling bales spreads like a contagion. It degrades nearby dry bales and forces paper mills to downgrade entire loads. Material that would be fully recoverable becomes second-grade furnish.

After environmental impacts, regulations become critical. European Regulation (EU) 2025/40, the Packaging and Packaging Waste Regulation (PPWR), entered into force on 11 February 2025. The full application is due on 12 August 2026 [26]. The regulation requires minimising packaging weight and volume while maintaining functionality. This is both right and necessary. Yet it creates a paradox.

The industry now uses lighter grammages, finer flute profiles, and narrower glue tabs. By reducing material per pack, each step tightens the margin for error. For instance, lighter liners are more brittle and crack during creasing, while lighter fluting media cause more washboarding. Additionally, narrower glue tabs mean less bonding area and less backup if a glue point fails.

The objective, therefore, cannot simply be less material. The objective must be smarter material choices and smarter processes. We need fewer grams per pack, but with processes that reliably deliver dimensional stability.

Smarter processes require a clear understanding of the challenge. That starts with a key distinction: which deviations can a line manager fix by adjustment, and which cannot be eliminated by any machine setting. This difference will run through the rest of this book.

1.5 Variation Is the Enemy, Not the Individual Deviation

If a machine systematically creases 1.5 millimetres too wide, that is a problem which can be resolved through adjustment. A technician corrects the setting, and the deviation is eliminated. Systematic errors of this kind are manageable.

What is not manageable is variation. If every tenth box has a different spring-back force because the crease line lands sometimes on a flute tip, sometimes in a valley and sometimes on a flute shoulder, no machine setting can help, because there is no single value to correct.

Nagasawa and Komiyama (2012) measured this effect with precision[29]. They defined the normalised deviation of the crease position from the flute tip and found that the greater the deviation, the greater the scatter in peak bending moment. The implication is clear: it is not the mean crease quality that determines box quality, but the variance in crease quality.

Nica et al. (2020) confirm this finding for inline creasing directly at the corrugator [30]. Because the crease is applied in the machine direction, the phase relationship between the creasing tool and the flute geometry varies both across the web width and over time. The result is an unavoidable scatter in the point of contact – tip, shoulder or valley – and therefore an unavoidable scatter in crease quality.

This insight is fundamental. Conventional process optimisation – the search for the correct setting value – reaches its limits when the underlying problem is variance. What is needed are methods that reduce the consequences of that variance rather than attempting to eliminate its causes. The two sections that follow introduce the principles behind those methods.

1.6 A Lesson from Aerospace: Designing for Damage Tolerance

To illustrate the principle behind the first of these solutions, it helps to look at another industry – one where the failure of a joint costs not complaints but lives. The aviation industry long ago accepted that micro-cracks in metallic structures cannot be entirely prevented over a service life of decades. What can be engineered is not their absence but their containment: fuselage skins are subdivided by stringers, frames and doublers that act as crack arrestors, so a propagating crack loses its energy at the next structural boundary and the aircraft loses a panel rather than its integrity.

The regulatory language for this is damage tolerance: a structure is certified not because it is defect-free, but because it can carry its operational loads with the defects it will realistically contain. Translated to a corrugated box, a continuous bead of adhesive along the manufacturer's joint is a fuselage skin without stringers – maximum strength on a perfect day, but vulnerable to a single propagating failure. Short, offset segments of adhesive – stitching – introduce the same crack-arresting logic at the glue line. The underlying fracture mechanics and the trade-off between peak strength and tolerance are developed in detail in Chapter 10. For now, the insight that matters is simpler: the perfect glue line does not exist, and what matters is what happens when the real one falls short of perfect.

1.7 A Second Analogy: Medicine Rather Than Surgery

This book's second core idea is a medical analogy. The corrugated industry typically responds to problems with more force: extra creasing pressure for cleaner folds, more adhesive for stronger bonds, and higher nip pressure for better contact. When results fall short, intensity is simply increased.

The problem is that every such intervention carries side effects. Higher creasing pressure results in greater crushing, causing permanent damage to the flute structure. Garbowski et al. (2021) show that at 70 per cent crush, bending stiffness falls by up to 50 per cent and shear stiffness drops disproportionately [4]. More adhesive extends open time and introduces additional moisture into the substrate, which in lightweight liners leads to saturation. Higher nip pressure displaces the adhesive from precisely the locations where it is needed.

The alternative is pharmacological: targeted, measured and local. Waterscoring – the localised application of a surfactant solution along the crease line – follows this principle exactly. It does not modify the machine; it modifies the material at the critical point. It reduces bending stiffness along the crease line, making the fold more precise and lowering spring-back forces. It increases the liner's elasticity, preventing cracking. And it homogenises crease quality across varying contact positions between the creasing tool and the flute geometry.

As with any medication, the substance itself does not determine the outcome: the dose does. Too much moisture destabilises the material. Too little has no effect. The skill lies in calibrating the right dose to the liner type, flute profile and ambient conditions.

For now, remember: smarter intervention, not stronger effort, is key.

1.8 Thread Running Through This Book

The central argument of this book is that the most serious threat to corrugated packaging lies in uncontrolled variation, not in individual defects. Addressing this risk requires not stronger interventions but smarter, more targeted ways to condition the process and material.

The following chapters provide the evidence. Let us begin where these problems originate.

Chapter 2: Where Defects Begin: Washboarding, Warp, Striping and Crushing as the Corrugator's Legacy

Introduction card

Four corrugator-side defects shape almost every downstream step: washboarding, warp, striping, and crushing. This chapter explains where they originate, how they travel through production, and why the folder-gluer can only manage their effects, not remove their causes. It is essential for readers who need to understand which upstream conditions later become folding, bonding, stacking, or retail problems.

"*Which defects originate at the corrugator, what causes them, and what consequences do they produce at the downstream stations that the corrugator team rarely sees?*"

Economic impact: Corrugator-side defects create distinct downstream costs. Washboarding, caused by excess starch and differential surface deformation, reduces print quality when amplitude exceeds 50 microns. Above 100 microns, local unbonded valleys can contribute to joint failures that first appear at the case erector. Warp, caused by moisture differences between liners, can create spring-back that opens the manufacturer's joint. Warp stores elastic energy and can add peel stress at the manufacturer's joint once the blank is forced flat and then released. Striping and flute-height variation across the web create inconsistent compressive strength and can lead to box rejections at retail. Crushing reduces stiffness and strength without always leaving a visible surface signature.

Ecological impact: Material reduction increases sensitivity to all four defects. Lighter facings have less moisture buffering, narrower flute-height tolerance, and lower starch-bond margins. They are therefore more vulnerable to warp, striping, washboarding, and crushing. The practical response is better dosing, careful drying, and less mechanical damage.

Operational efficiency: Crushing during corrugating, printing, transfer, or slitting reduces shear stiffness and creates crease-quality variation that the folder-gluer cannot correct. Pressure-measurement film at each nip point helps identify the source. In many cases, correction requires a 10 to 20 per cent reduction in nip pressure rather than new hardware. Once measured, many plants find the primary problem at the slitter or another transfer point, not necessarily at the dryer.

Quality: Each defect has a measurable parameter or diagnostic signature. Washboarding should remain below 50 microns. Above 100 microns, it becomes critical. Each defect has a measurable parameter or diagnostic signature. Washboarding is measured as surface amplitude. Warp appears as millimetre-scale board curvature. Striping appears as periodic ink-density variation. Crushing requires special attention because the board can appear acceptable while its load-bearing capacity has already been reduced. Flute-height variation across the web should stay within ±2 per cent of nominal. Crushing requires special attention because the board can appear acceptable while its load-bearing capacity has already been reduced.

Limitations: Corrugator defects cannot be corrected at the folder-gluer. Persistent washboarding above 100 microns, warp above 8 mm, or significant crushing requires upstream intervention. Downstream conditioning can mitigate some effects, but it cannot solve their causes. Responsibility for correction remains with the corrugator and upstream handling process.

Read in full: corrugator process engineers, plant managers, quality managers; folder-gluer operators dealing with persistent fishtailing or pop-opens that have no obvious downstream cause.

Card sufficient: readers whose responsibility starts well downstream of the corrugator and who need to understand what arrives at their station.

Prerequisite: Chapter 1.

Follow-up: Chapter 3, where defects converge at the manufacturer's joint.

2.1 Four Defects, Three Orders of Magnitude

No corrugator in the world delivers a perfect board. What it delivers is a compromise between speed, adhesive quantity, temperature and moisture – and that compromise leaves its mark. Four of those marks shape everything that follows at the flexo folder-gluer: washboarding, warp, striping, and, less visibly, crushing.

Three of them differ in scale by a factor of a thousand, yet act together on the quality of the finished pack. Beck and Fischerauer (2022) have shown that they can be separated from one another with mathematical clarity[1].

Warp follows a smooth surface function that describes the large-scale curvature of the entire blank, with typical amplitudes of 2–3 mm. Washboarding is the periodic residual component that follows the flute structure: a modulation of the liner surface with an amplitude of 10 to 100 µm.

Striping, finally, is not a geometric defect in the strict sense, but an optical signature: a periodic variation in ink density in post-print.

Kirchhoff plate theory provides the framework for this separation: warp as the low-frequency bending of the board as a whole, washboarding as a high-frequency, flute-periodic fluctuation. The distinction is not academic. It is operationally decisive because the three defects originate at different points in the process chain, require different countermeasures, and, at the folder-gluer, combine into a superimposed, compounded problem.

The fourth defect—crushing—does not fit the geometric hierarchy. It has no characteristic scale because, by design, it is invisible. Crushing is a permanent loss of flute height and mechanical integrity, leaving no surface signature. Section 2.7 treats it in detail, but it belongs in the opening inventory: what the folder-gluer inherits from the corrugator is not three defects, but four.

2.2 Washboarding: The Periodic Surface Undulation

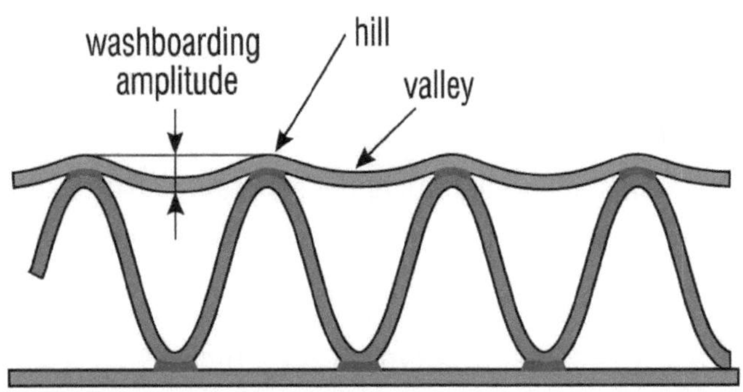

Figure 1 Washboarding

Washboarding is the most widely discussed and most thoroughly researched form of surface disturbance in corrugated board. The name aptly describes the phenomenon: the linear surface exhibits a washboard-like relief that follows the underlying flute structure. The amplitude is barely visible to the naked eye, on the order of 10–100 micrometres, yet its effects on contact printing, bonding, and downstream converting are considerable.

The Root Cause: Adhesive Add-On at the Corrugator

The doctoral thesis by Wendler (2006, RMIT University, Melbourne) quantified what the industry had long known qualitatively: adhesive add-on at the corrugator is the dominant influencing factor[19].

Wendler used digital image profilometry and Fourier analysis to measure the surface undulation of the liner with frequency resolution. The relationship between starch adhesive add-on and washboarding amplitude is almost linear. Every gram of excess starch per square metre amplifies the deformation. The implication for lighter paper weights is direct: less fibre mass means higher washboarding amplitudes and poorer conditions at the folder-gluer.

There is, however, a counterpoint that complicates any simple call for less adhesive. Popil and Schaepe (2007, Georgia Institute of Technology) showed, through nonlinear finite-element modelling and experiments, that higher adhesive levels at the corrugator substantially increase the Edge Crush Test (ECT) value[13].

The mechanism is not the stiffness of the dried adhesive film itself, whose contribution to the effective board modulus is only around three per cent, but the inhibition of interflute buckling: the adhesive constrains the liner panels between the flute tips and prevents them from buckling out of plane under compressive load. At a 4.6-fold increase in glue volume, ECT rose by up to 30 per cent. The corrugator operator, therefore, faces a genuine dilemma: reducing adhesive add-on lowers washboarding but simultaneously removes the buckling restraint that protects compressive strength.

Differential Shrinkage and Flute-Tip Densification

Figure 2 Different degrees and manifestations of washboarding

To understand why the adhesive add-on is so decisive, it helps to look at the micro level. The starch adhesive bonds the liner to the flute tips of the medium. At these bonding points, the liner is fixed; between them, spanning the flute valleys, it is free. As the adhesive dries, it shrinks and pulls the liner downward at the tips. The more adhesive, the greater the tensile force, and the more pronounced the indentation. The liner bows downward between the tips: this is washboarding.

This effect is further amplified by the paper's hygroexpansion. Vonk et al. (2020) measured the swelling of individual cellulose fibres with a precision of 1×10^{-4}[17]. Fibres swell transversely to the fibre direction by 20 to 30 per cent more than they do longitudinally. This anisotropy carries through to the liner. When moisture is absorbed, the liner expands between the fixed bonding points at the flute tips. Because it is pinned, the expansion has nowhere to go but into the valleys, so the washboarding deepens further during storage.

The starch adhesive, therefore, creates zones in the liner that are less porous at the flute tips, while the valleys remain fast-absorbing. Where the adhesive is most needed later at the MJ, it flows longest. Ramos-Gonzalez (2015) treats this 'flute-tip densification' in depth and shows that it is not a side effect but a permanent structural property of the board[64]. Its consequences for bonding at the manufacturer's joint follow in Chapter 3.

Bonding Defects as Washboarding Amplifiers

Johnson and Popil (2015) classified five failure modes of the corrugated board adhesive bond. These range from adhesive failure (the most brittle) through failure at the medium-to-adhesive interface to failure within the liner itself (the strongest bond)[9]. Their thermoelastic stress analysis using infrared imaging reveals the stress distribution around bonding points. They show how defects—insufficient corrugator temperature producing white, under-gelatinised glue lines, or excessive temperature producing brittle bonds without adhesive penetration—alter local stiffness and amplify the washboarding signature.

2.3 Moisture as a Time Bomb

Washboarding is not a static defect; it intensifies over time, driven by moisture. Tuomainen et al. (2023) were the first to use magnetic resonance imaging to show how water vapour penetrates corrugated board at high spatial resolution[16]. Absorption occurs mainly in the gas phase via the air channels within the flute structure, not through the paper matrix itself. The fluting medium absorbs moisture faster than the liners. Bronlund et al. (2013) confirm, through finite-element modelling, that most moisture transport occurs via the air phase [2].

Building on this, Cornaggia et al. (2023) show by numerical homogenisation that the individual plies of corrugated board respond differently to moisture[3]. The fluting medium is the most sensitive. Modzelewska (2006) independently confirmed this finding[10].

The practical consequence is significant. Between the corrugator and the folder-gluer, hours or even days may pass. During that time, moisture continues to act on the board. The washboarding amplitude leaving the corrugator may be more pronounced by the time the board reaches the folder-gluer than it was at the point of manufacture – especially in non-climate-controlled storage or when the board is brought in from a cool warehouse. In parallel, bulk mechanical properties degrade. Rather than scatter the numbers through the prose, the table below consolidates what the literature reports:

Humidity condition	Measured effect	Material / test	Source
50–90 % RH	Up to 22 % loss in compressive strength	Corrugated board under compressive load	Szewczyk & Głowacki 2014 [15]
33 → 93 % RH swing	ECT drops by 2 to 3 N/m	Single-wall board	Modzelewska 2006 [10]
Variable RH	0.4 to 1.2 % hygroexpansion in the liners	Liner paper, B-/C-flute board	Modzelewska 2006 [10]
Variable RH	Markedly higher expansion than in the liners (no single closed-form figure)	Fluting medium	Modzelewska 2006 [10]; Cornaggia et al. 2023 [3]

Taken together, the board softens, the washboarding deepens, and the MJ bond begins under increasingly unfavourable conditions. Moreover, storage after the corrugator is not a pause but a continuation of the process.

2.4 The Moisture Window: Why Storage Time Is a Process Step

Figure 3 Variable storage times occur between corrugator production and converting

Moisture not only amplifies washboarding; it also determines whether a blank is processable at all. Garg et al. (2016) documented the moisture profile during production: from 6 to 8 per cent at the board infeed, through 8 to 11 per cent after flute formation, to 11 to 23 per cent after lamination[6]. After four hours of drying at 45 per cent RH and 22 °C, it settles at 10 to 14 per cent.

Within that range, the board has a narrow usable window. Consolidating the five characteristic bands into a single three-column view makes the operational decision at the FFG infeed explicit:

Moisture content	Material behaviour	Operational decision
< 8 %	Brittle; cracks during creasing	Do not convert; return to conditioning
10–12 %	Optimum processability for printing and converting	Target window for FFG infeed; convert now
12–14 %	Softening begins; strength starts to fall	Monitor; shorten storage time before converting
14–18 %	Area-wide expansion; print register accuracy lost	Postpone converting; dry first
19–23 %	Severe expansion on creasing, contraction on drying	Reject; permanent deformation is unavoidable

The optimum print and converting window is 10-12 per cent. Below 8 per cent, the board cracks at the crease. Above 14 per cent, the material softens, and print registration suffers; between 14 and 18 per cent, the paper expands across its full area and contracts again on drying, with dimensional accuracy and register precision both lost.

The implication is the same as in Section 2.3: the time between corrugator and folder-gluer is not neutral. This is a process step in its own right. It directly governs the quality of the finished pack.

2.5 Warp: Large-Scale Board Curvature

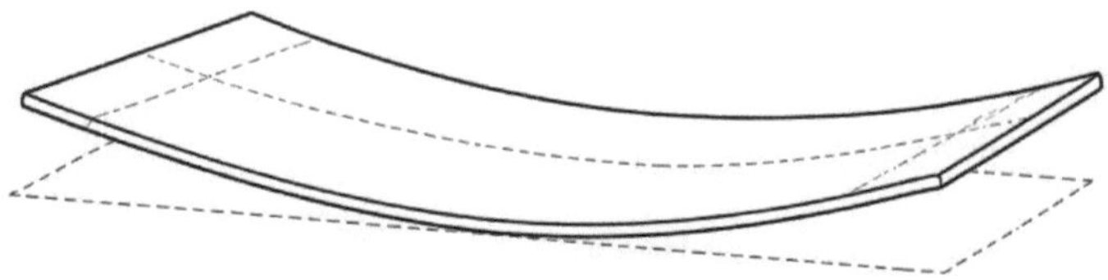

Figure 4 Wrap (represented here as a bow-shaped configuration)

Where washboarding operates at the micrometre scale, warp operates at the millimetre scale. Warp is a macroscopic deformation of the entire blank, caused by internal stresses during drying and cooling at the corrugator. Typical amplitudes are in the range of 2 to 3 mm, measured across the sheet's diagonal – a thousand times greater than washboarding – and where washboarding is a local modulation, warp affects the overall geometry of the blank.

The Three Basic Forms of Warp

Bow (longitudinal curvature): the blank curves along the flute direction. At the MJ, this means the glue tab is bowed along its length, either upward or downward, and the distance between the applicator and the substrate varies continuously.

Curl (transverse curvature): the blank curves across the flute direction. Because the glue tab is narrow (typically 25 to 35 mm), the curl effect on the tab itself is limited, but it acts on the overall packaging geometry during joining.

Twist (torsional distortion): the corners of the blank do not lie in a single plane. Twist is particularly critical for MJ bonding because it introduces a relative angular deviation between the flute directions of the two mating surfaces. Wade et al. (2023) show that even at a flute angle deviation of 5 to 10 degrees, edge crush resistance falls by 5 to 10 per cent, and at 60 degrees by 41 per cent [18].

Warp form	Adhesive application	Flute alignment	Peel / opening risk
Bow	Gap varies along tab length	Uneven pressure distribution (centre vs. ends)	High: lever effect at tab ends
Curl	Limited effect on narrow tab	Overall packaging geometry alters nip pressure distribution	Moderate: acts through overall geometry
Twist	Combination of bow and curl	Flute angle deviation produces moiré pattern	Very high: asymmetric opening force

The Stored Spring-Back Impulse

Warp is not a harmless visual deviation. The blank stores elastic energy. During joining at the folder-gluer, it is forced flat by drive belts and rollers; the glue line is closed under this enforced contact. The moment the nip pressure is released on exit from the pressing section – and at the latest in the delivery stacker – the board attempts to return to its equilibrium position.

This spring-back force acts to peel the glue line open, much like peeling a sticker off, or to hinge it open on one side like a book. Both loading modes are far more dangerous to a glue line than uniform tensile stress, because the force concentrates along a narrow line rather than being distributed across the full bonding area. At the tab edges, where warp generates the greatest lever action, peel dominates. Across the tab surface, a one-sided opening occurs when the spring-back force is distributed asymmetrically – typically the case with twist. A uniformly distributed area load virtually never occurs in practice, because spring-back motion always includes an opening component perpendicular to the bonding surface. Chapter 3 systematically classifies and quantifies these loading modes.

2.6 Striping: When the Flute Structure Shows Through

Figure 5 Striping

Striping is the third member of the group and behaves differently from both washboarding and warp. It is not a geometric defect but an optical one: a periodic variation in ink density in post-print that follows the flute structure. Holmvall and Uesaka (2010) have clarified the mechanism with precision[7]. The nip pressure in the printing unit varies periodically: pressure peaks over the flute tips, pressure minima over the valleys. That pressure variation translates directly into a change in ink density. One finding in particular tends to surprise practitioners: print gloss is almost entirely independent of nip pressure. Striping is a pure ink-density phenomenon, not a gloss phenomenon.

A second, optical-chemical dimension reinforces the effect. At the flute tips, the liner has been pre-impregnated and compacted during corrugator bonding; at the valleys, it remains porous and open. Where the substrate is denser and less absorbent, ink sits higher on the surface; where it is porous, ink penetrates more rapidly. This differential absorption reinforces the periodic density variation that nip pressure alone would produce and compounds the visual striping. The underlying flute-tip densification is a permanent structural property of the board[64]; Chapter 3 returns to it as one of the three principal mechanisms of uneven bonding at the manufacturer's joint.

Post-Print Technology Matters: Flexo versus Single-Pass Inkjet

The mechanism described above was first investigated for flexo post-print, where the topographical signature is mediated by contact between the plate and the board. Single-pass inkjet post-print, which is gaining ground rapidly in short-run and late-stage-customisation applications, is a contactless process and therefore avoids nip pressure altogether. What it cannot avoid is the optical-chemical component: the differential absorption between the densified tip and the porous valley persists, and the higher drop-placement resolution of inkjet makes topographical non-uniformity even more visible in mid-tones and fine graphics.

The practical consequence is that moving from flexo to digital post-print does not eliminate striping; it shifts its origin from the mechanical to the absorptive side. Targeted surface treatment of the liner and tighter control of adhesive add-on at the corrugator, therefore, become more, not less, important as digital post-print share grows. This deserves a fuller treatment than is possible here and is picked up again in Chapter 11.

The Fundamental Trade-off

The obvious countermeasure in flexo – increasing nip pressure (squeeze) – reduces striping but carries a serious risk: above approximately 0.23 MPa, damage to the flute structure begins. Garbowski et al. (2021) have systematically demonstrated that crushing degrades the mechanical properties of corrugated board [4]. The ECT value drops by 1.5-2.0 N/mm, and approximately half of the deformation is plastic and permanent. The full extent of crushing damage is discussed in Section 2.7.

This is a trade-off that cannot be resolved, only managed. The nip pressure required to achieve saturated ink density is close to the board's damage threshold. Maximising print quality risks mechanical integrity; protecting structural performance means accepting print variation. For the folder-gluer, the board that has already been over-crushed at the printer presents a compounded problem: reduced bending stiffness, altered creasing behaviour and a flute structure with diminished capacity to carry compressive loads on the pallet.

2.7 Crushing: The Hidden Driver of Dimensional Change

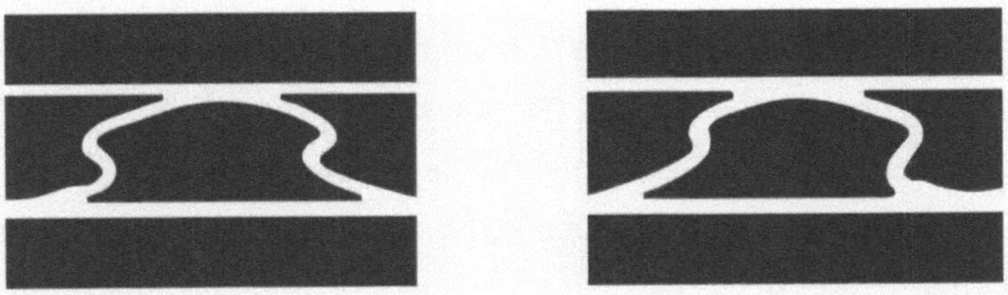

Figure 6 Typical crushing patterns

Alongside washboarding, warp and striping, there is a fourth defect type that is frequently overlooked precisely because it is invisible: crushing. Garbowski et al. (2021) conducted a systematic investigation into how the mechanical properties of corrugated board deteriorate during compression during converting [4]. The results are alarming. Bending stiffness decreases linearly with increasing crush and can drop by up to 50 per cent at a thickness reduction of 70 per cent. Torsional stiffness collapses at just 40% crush. The ECT value drops by 1.5–2.0 N/mm. Approximately half of the deformation recovers elastically; the other half is plastic and permanent.

Rogalka et al. (2024) developed a CNN-based system capable of detecting crushed corrugated board in real time [14]. One finding of their work deserves particular attention: in some cases, crushed B- and C-flute board is misclassified as E-flute because compression blurs the boundaries between flute types. When a quality system identifies E-flute, it may in fact be looking at a damaged B- or C-flute structure.

Jamsari (2020, Massey University) adds a further dimension [8]. Using laser-cut cross-sections and digital image analysis, he developed a method to quantify flute profile damage using a 'Similarity Factor' that compares the actual flute geometry to an undamaged reference. Two findings are particularly relevant. First, torsional behaviour shows the highest sensitivity to crushing even at low compression levels, independently confirming Garbowski's observation. Second, and more surprisingly, the flute geometry itself partially recovers after crushing, yet the strength loss persists. Jamsari attributes this to fibre-level damage within the constituent papers, damage that is invisible even under microscopic inspection of the flute profile. This explains why crushing is so insidious: the board may look geometrically intact, yet its load-bearing capacity has been permanently degraded.

For the folder-gluer, crushing results in altered creasing behaviour, reduced bending stiffness at the fold edge, and a flute structure with diminished load-bearing capacity on the pallet. The defect is invisible in the finished pack, but its effects are not.

2.8 Superposition: Why One Plus One Does Not Equal Two

None of these defects occurs in isolation. A real blank leaving the corrugator and arriving at the folder-gluer hours or days later carries all of them simultaneously: washboarding at the micro level, warp at the macro level, possibly crushing damage from the printing unit, and a moisture content that has shifted since production.

The combination is more than the sum of its parts. Washboarding sits on top of the warped surface: an applicator designed to compensate for washboarding must also contend with the large-scale deviation introduced by warp. Moisture amplifies both – it increases the washboarding amplitude and the warp spring-back force. Crushing alters the geometry on which both washboarding and warp act. And stripping signals that the printing unit may already have introduced damage, weakening the flute structure.

This superposition explains why setting up the folder-gluer is not a one-time task but an ongoing compromise. Incoming board quality varies from reel to reel, from pallet to pallet, and from morning to afternoon. What began at the corrugator as an acceptable compromise compounds at the folder-gluer into a problem that conventional setting optimisation struggles to contain.

2.9 Measurement Methods: Making the Invisible Visible

The good news is that all of the defects described are measurable. Measurement technology has advanced considerably over the past decade, and the combination of optical sensing and machine learning is opening up possibilities that were previously unthinkable.

Method	Principle	Practical guidance	Source
Digital profilometry	Optical surface topography with Fourier analysis, 10–100 µm resolution	Lab method; use for root-cause analysis of washboarding complaints	Wendler 2006 [19]
Laser triangulation	Inline real-time scanning of the board surface, < 50 µm	Inline capable; install after corrugator for continuous warp and washboarding monitoring	Beck & Fischerauer 2022 [1]
Iodine staining (TAPPI T 610)	Visualisation of starch distribution at flute tips	Simple shop-floor test; verifies corrugator adhesive distribution	Johnson & Popil 2015 [9]
IR thermography	Thermoelastic stress analysis of bonding zones	Lab method; reveals stress concentrations invisible to visual inspection	Johnson & Popil 2015 [9]
MRI imaging	Spatial moisture distribution in cross-section, < 1 mm	Research tool; quantifies moisture penetration paths and validates FE models	Tuomainen et al. 2023 [16]

CNN classification	AI-based detection of flute types and deformation, 99 % accuracy	Inline capable; alerts operators to crushing damage before converting	Rogalka et al. 2024 [14]
FT-NIR spectroscopy	Non-destructive classification of paper grades	Incoming-goods inspection; verifies paper grade matches specification	Pensold & Plew 2003 — see bibliography note

Garbowski (2024) describes how AI systems in production can detect and optimise, in real time, the relationships among adhesive add-on, temperature, moisture, and the resulting board properties [5]. The challenge lies less in the measurement technology itself than in integration: inline measurements at the corrugator must be linked to the process parameters at the folder-gluer, so that the information arrives where it is actually needed.

2.10 The Corrugator's Legacy: What the Folder-Gluer Inherits

The corrugator leaves the folder-gluer with a material that carries four inherited burdens. First, a modulated surface – washboarding – that influences adhesive application, the contact area during joining and the strike-in behaviour of the MJ adhesive. The amplitude is a function of the adhesive add-on at the corrugator and increases further through moisture absorption during storage.

Second, a macroscopic curvature – warp – that stores elastic energy and, after joining, acts on the still-fresh glue line as a peel and opening force.

Third, a differential porosity – flute-tip densification – that alters the strike-in behaviour of the MJ adhesive depending on location and impairs the development of bond strength across the glue line.

Fourth, a potentially pre-damaged flute structure – crushing – that reduces bending stiffness and edge crush resistance without any visible external indication.

None of these inherited burdens can be eliminated at the folder-gluer. They can only be compensated, mitigated, or, at best, managed in a way that keeps their effects under control. That is precisely the subject of the next chapter: how the folder-gluer addresses this legacy, and why the manufacturer's joint is the point at which all of these defects converge into a single, critical point of failure.

2.11 Chapter 2 at a Glance

Defect overview

Defect	Scale	Primary cause	Effect on MJ	Key reference
Washboarding	10–100 µm	Adhesive add-on at corrugator	Differential porosity; uneven adhesive strike-in	Wendler 2006 [19]; Netz 1998 [11]
Warp	2–3 mm	Internal stress from drying/cooling	Spring-back peel force on fresh glue line	Beck & Fischerauer 2022 [1]
Striping	Optical	Nip pressure variation + differential absorption at flute tips	Signals potential crush damage and flute-tip densification	Holmvall & Uesaka 2010 [7]
Crushing	Invisible	Excessive nip pressure in printing / converting	Reduced bending stiffness; altered creasing	Garbowski et al. 2021 [4]

Five failure modes of the corrugated board adhesive bond (Johnson & Popil, 2015)

Mode	Description	Assessment
1	Adhesive detaches completely from the substrate	Most brittle mode; weak bond
2	Failure at the contact surface with the fluting medium	Insufficient adhesive penetration
3	Flute tips tear away (decapping)	Fluting medium is the weak point
4	Failure between adhesive and liner	Moderate bond
5	Paper ruptures; bond stronger than substrate (fibre tear)	Strongest bond

A quote to carry forward

When a quality system identifies E-flute, it may in fact be looking at a damaged B- or C-flute structure. Crushing is the only defect that systematically misrepresents the material it has changed. Chapter 2: Where Defects Begin: Washboarding, Warp, Striping and Crushing as the Corrugator's Legacy

Part II -
The Consequences, or Pathology

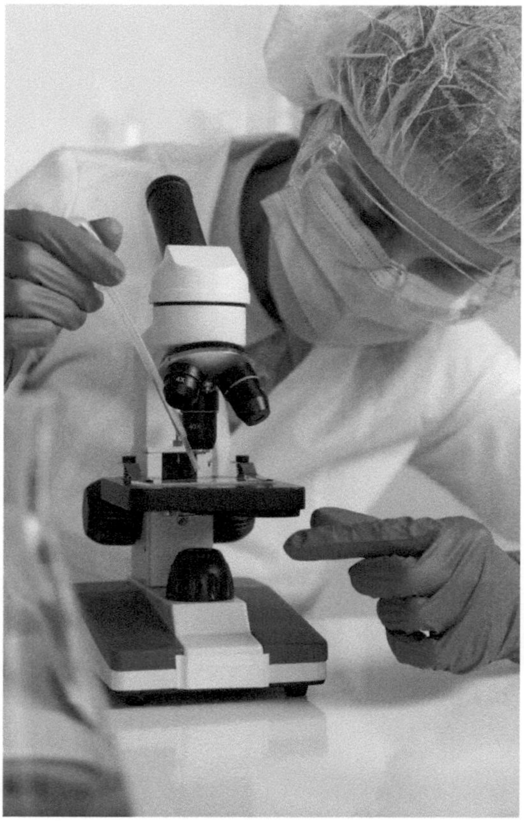

Chapter 3: Where Defects Converge: The Manufacturer's Joint as the Focal Point of Failure

Introduction card

A box can pass laboratory tests and still fail at the manufacturer's joint. This chapter explains why: the joint is not mainly loaded in the mode that BCT measures, but in peel and cleavage during the green-bond phase. It is central for readers who need to connect adhesive behaviour, crease mechanics, liner properties, and field failures.

"Why is the manufacturer's joint the failure point even when no individual upstream parameter is out of specification, and why can a box that passes BCT in the laboratory still pop open in the supply chain?"

Quality: The manufacturer's joint is exposed to three load modes: shear, peel, and cleavage. BCT mainly represents shear, where PVAc performs well and may reach strengths close to the substrate. In practice, the joint is often loaded in peel and cleavage after folding. These are the weakest load modes for PVAc. Spring-back at the crease creates the opening force, so the field failure occurs where the adhesive is least capable of resisting it.

Operational efficiency: The first seconds and minutes after joining are critical. In this period, green strength must resist spring-back before full film formation has developed. During this green-bond window, PVAc has reached only a small share of its final strength, while spring-back is already acting at full force. The joint can therefore open before the adhesive has set. The case erector or filling line then sees the defect as a pop-open or fishtailing-related jam.

Economic impact: Adhesive consumption at typical packing-line settings is a major annual cost. At 4 to 5 grams per bonded side and PVAc prices of about EUR 2.0 to 3.0 per kilogram, adhesive becomes a six-figure cost line in mid-volume FMCG and a seven-figure cost line in high-volume beverage. The main driver is not final bond strength, but the holding force needed during the green-bond phase.

Ecological impact: Reducing adhesive use lowers the embedded carbon linked to PVAc production. Preventing joint failure has a larger ecological effect, because it also protects the product inside the pack. These two savings are separate and cumulative.

Limitations: This chapter explains failure mechanics. It does not yet introduce the interventions. Readers looking for solutions should continue to Part IV, especially Chapters 10 and 11.

Read in full: process engineers, adhesive specifiers, anyone making capex decisions on folder-gluer or adhesive systems.

Card sufficient: executives reviewing the cost-of-failure argument; sales engineers explaining why BCT does not predict field failure.

Prerequisite: Chapter 2.

Follow-up: Chapter 4, the downstream cascade.

3.1 Five Phases, Five Risk Points

The defects described in Chapter 2 occur at the flexo folder-gluer, the critical process step where the manufacturer's joint (MJ) is bonded. The corrugated blank, already creased and slotted, is closed into a three-dimensional box at the glue line. Joint failure results in box failure.

The process can be divided into five phases. Each carries its own specific risks. Despite differences, they share a theme: the geometry of corrugated board creates asymmetry. Defects from the corrugator amplify, rather than create, that asymmetry.

Even corrugated board without washboarding or warp has a liner that freely spans valleys between flute tips. Under pressure—during adhesive application, joining, or bundling—the liner deflects more in valleys than at flute tips. Thinner liners accentuate this effect. Nip pressure, adhesive distribution, and contact area are inherently non-uniform. Washboarding and warp intensify, but do not cause, this geometric asymmetry.

Phase 1: Adhesive Application

At the adhesive application unit, two to three beads of PVAc (polyvinyl acetate) dispersion adhesive are applied to the glue tab, typically at 50 per cent solids content and with a viscosity of 400 to 2,000 mPa·s. Each bead is 2 to 3 mm wide and initially forms a domed profile.

This is where the adhesive first meets the corrugator's legacy. Contact systems, such as roller or nozzle application, touch the board surface directly. On a washboard surface, the nozzle mainly contacts the peaks and wipes adhesive from them. More adhesive remains in the valleys. Patent EP 2 279 856 uses reaction-force sensors to adaptively regulate nip pressure[25]. Under severe washboarding and high machine speeds, the control dynamics may not keep pace.

Even without measurable washboarding, contact pressure varies. The liner deflects between flute tips, so nozzles or rollers press harder at support points and less in unsupported spans. For lightweight liners with recycled fibre near 100 g/m^2, this deflection is noticeable. Washboarding increases the effect, but the geometric asymmetry is the root cause.

A further problem relates to crush sensitivity (Chapter 2). Contact systems can damage the liner surface under elevated nip pressure and are prone to smearing, especially at higher speeds and with varying substrate calliper. These disadvantages have led non-contact systems to become the near-universal choice in modern flexo folder-gluers. Non-contact (free-jet) application does not touch the substrate; the applied quantity is independent of surface topography, and a bead landing on a peak or in a valley affects position, not the amount deposited. Warp alters the standoff distance between the nozzle and the substrate and, under pronounced curvature, can affect the point of impingement and the bead geometry.

Phase 2: Open Time: The Race Begins

After the application, a race begins. The adhesive loses water through strike-in to the substrate. Polymer particles concentrate and form a film. At the same time, gravity acts on the domed bead, causing lateral spreading. On washboarding peaks, strike-in is slower because the liner has been pre-treated with corrugator starch adhesive (flute-tip densification); in the valleys, the liner is untreated and absorbs rapidly [64].

The result is paradoxical. Adhesive is most needed on the peaks, where viscosity remains lowest, and flow is highest. In the valleys, the adhesive quickly becomes immobile, despite less contact with the opposing surface. More pronounced washboarding intensifies this effect.

Phase 3: Joining: The Critical Moment

Joining presses the adhesive-coated tab to the opposing surface. This moment is even more critical for adhesive distribution than Phase 2.

Even without washboarding, flute geometry creates uneven contact. The liner deflects in the valleys under pressing force, while flute tips act as load-bearing ridges. Effective contact between the two surfaces is greater at the tip positions than in the valleys. With lightweight liners, valley deflection can leave the liner barely in contact with the adhesive. Where washboarding is present, this effect becomes much stronger. Surface topography compounds geometric compliance. Two washboarded surfaces then face each other, and their alignment determines the outcome.

Alignment	Effect on adhesive	Assessment
Tip to tip	Maximum contact pressure on the peaks. Adhesive is displaced from both sides into the valleys. Contact area is narrow.	Critical
Tip to valley	Peak of one surface presses into the valley of the other. Good geometric interlock; adhesive is displaced less.	Favourable
Skew run / moiré	Crossing flute directions due to skewed infeed. Tips and valleys superimpose into an interference pattern. Mixed contact conditions.	Variable

In a tip-to-tip configuration, pressing force is concentrated in narrow peak contact zones. Local surface pressure greatly exceeds nominal belt pressure. Flute damage begins at around 0.23 MPa[4]. Above this, adhesive displacement and permanent flute damage occur.

Phase 4: Setting Under Pressure

After joining, the belt pressure secures the bond. Distinguish between the two strength mechanisms in this phase, as they are often confused.

Green strength forms quickly. As the porous substrate draws water from the adhesive, the PVAc polymer particles move closer together. They form physical contacts through van der Waals forces and mechanical anchoring. This initial bond strength develops within seconds. It builds much faster on porous corrugated board than on non-porous substrates.

Film-formation strength takes longer to develop. Schneberger (1983)[32] and Pizzi and Mittal (2017)[31] describe four stages. The stages are: (1) discrete particles in an aqueous phase, (2) concentration and deformation, (3) initial strength from particle fusion, and (4) irreversible film formation with final bond strength. This process takes from minutes to hours.

The key question: can green strength withstand spring-back? On porous board, strike-in accelerates water removal five to tenfold compared to non-porous substrates (Washburn dynamics).

Washburn (1921) [81] provides the underlying capillary-flow model; Pizzi & Mittal (2017) [31] discuss the quantitative acceleration factors for porous cellulose substrates.

With a porous liner, moderate stiffness, and clean creasing, green strength withstands delivery stacker transfer. Dense kraftliner ($Cobb_{60}$ <30 g/m^2) delays strike-in and green strength. Stiff BC double- or triple-wall board, or poor creasing, leads to spring-back force exceeding green strength and peel stress localising along the glue line. In these cases, green strength is insufficient.

Time	Green strength	Film formation	Significance for the MJ
0 to 5 s	Building (water removal begins)	None	Belt pressure holds joint closed; particle contacts forming
5 to 30 s	Load-bearing on porous substrate	< 5 % of final strength	Delivery stacker: sufficient for lightweight board with good creasing; critical with stiff board
30 s to 5 min	Established	5 to 25 %	Bundle formation provides backup; film formation begins contributing
5 to 30 min	—	25 to 60 %	Handling strength
> 30 min	—	60 to 100 %	Full load capacity; final strength after 24 hours

The table's central point is clear: in the first seconds after joining, green strength—not film formation—carries the glue line. This green strength results from rapid water removal through the porous substrate. Under favourable conditions, green strength suffices. Under unfavourable conditions, the ratio of green strength to spring-back force determines success or failure.

Phase 5: The Delivery Stacker and Bundle Formation

After the pack leaves the belt and fans, nip pressure releases. Two forces act on the fresh glue: the blank seeks warp equilibrium, and crease-line stiffness opens the panels. The spring-back strength depends on board grade and crease quality. A deep crease on lightweight testliner generates less spring-back than an incomplete crease on stiff BC double-wall.

In practice, the majority of packs survive this moment as green strength proves sufficient. But the process is operating at the edge of its reserve. The most critical position is the topmost blank in the stacker: for a brief period, it lies exposed, unsupported by any weight. While the lower blanks are held together by the stack weight – so the PVAc dispersion can undergo film formation undisturbed – the topmost blank must rely on green strength alone.

Side joggers in the delivery stacker, often underestimated, align the bundle and, when dimensional deviations such as an overly wide pack occur, apply corrective force. This possible correction indicates the glue line remains plastic-viscous, allowing the adhesive film to flow and adapt to the relative movement of the bonded surfaces.

As long as the joggers produce only a brief alignment during the green-strength phase, the final bond strength is unaffected. Two scenarios become critical, however. The first is where the correction falls within the onset of film formation, from approximately 30 seconds onward. The polymer bridges that form between the PVAc particles are broken by mechanical displacement, and because the driving medium – water – has already largely migrated into the substrate, these bridges do not fully heal. Microcracks in the film are the result. The second arises when the joggers are permanently set too tightly, exerting a continuous lateral force on the joint. The partially cured adhesive film deforms under this sustained load through viscoelastic creep, weakening the bond. The slight bulging of the MJ area that becomes visible is the outward sign of this overstress. In practice, this can measurably reduce the final bond strength because the joint cures under a sustained load that does not align with its intended direction of stress.

A word of honest proportion. The best jogger setting in the world cannot rescue a pack whose dimensions are fundamentally out of tolerance. If the case that leaves the folding section is already several millimetres off width, the joggers either force a correction that damages the fresh film (scenario one above) or apply a continuous restraining force that weakens it (scenario two). Jogger tuning is a second-order lever – useful when incoming dimensions are within the band, powerless when they are not. The first-order lever maintains dimensional accuracy at the fold and the crease, as discussed in Chapters 7 and 11.

When the spring-back force exceeds the momentary green strength, the pack springs open. Pop-opening – the visible re-opening of the manufacturer's joint after the delivery stacker – is not an occasional defect; it is the visible sign that the ratio of spring-back force to green strength has crossed the critical threshold. Bundle size and dwell time are therefore not logistical details but process-critical parameters: they provide film formation with the time that green strength alone cannot provide.

A note on dwell time before strapping

Industry practice for standard single-wall grades on kraftliner or testliner is to allow the bundle to rest for approximately ten to twenty minutes before strapping. During this window, film formation progresses into the 25 to 60 per cent band (Section 4, table above), which is sufficient to absorb the residual spring-back force that was only partially balanced by green strength. For heavier grades (BC double-wall, triple-wall) or for cold, low-porosity substrates such as coated liners, a dwell time of thirty minutes or more is a more realistic target; for lightweight single-wall at warm, dry ambient conditions, ten minutes can be enough. These values are operational benchmarks, not universal specifications, and they interact with bundle height, ambient humidity and adhesive formulation. A simple rule of thumb: strap when the top blank of the bundle resists being lifted off by hand, and not before.

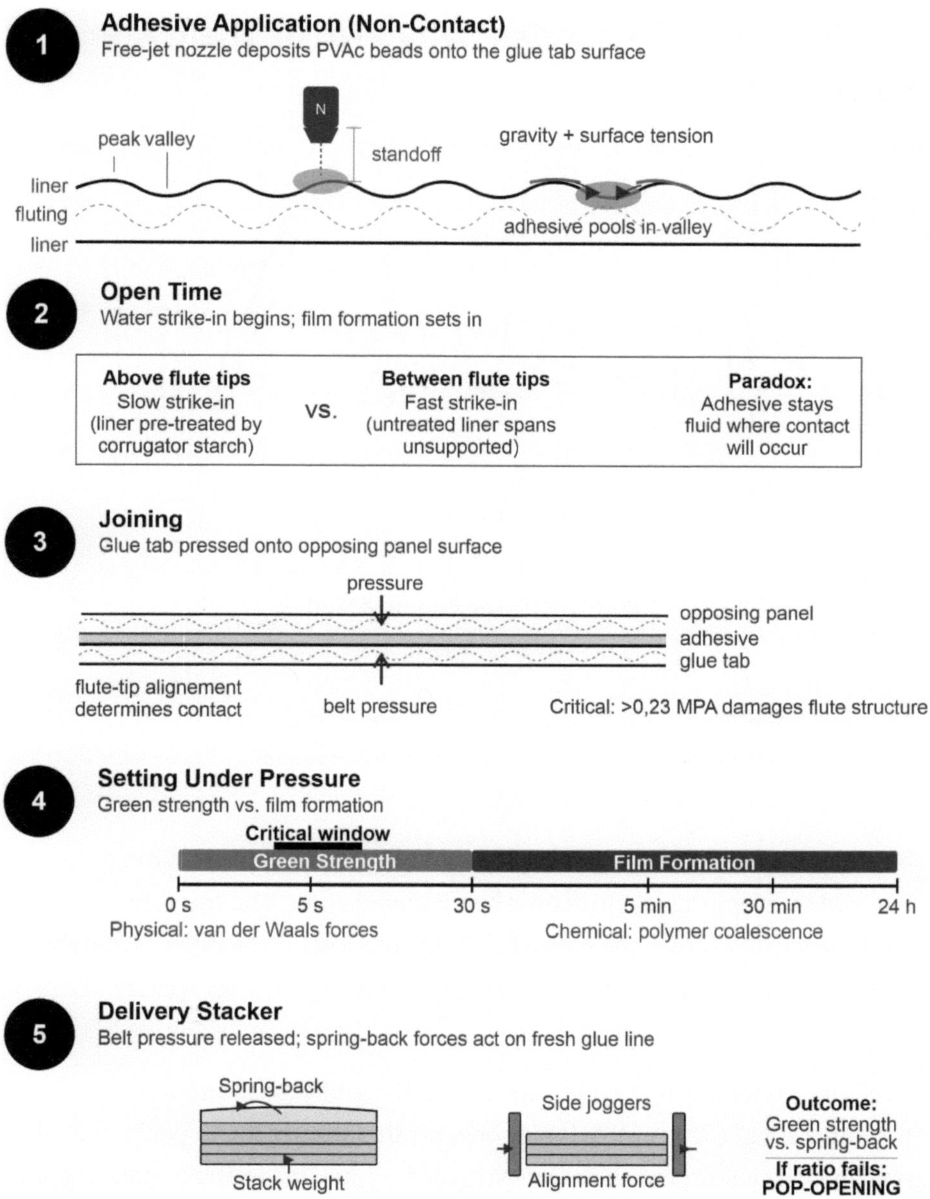

Figure 7 Five Phases, Five Risk Points

3.2 Three Loading Modes: Why Two of Them Are Dangerous

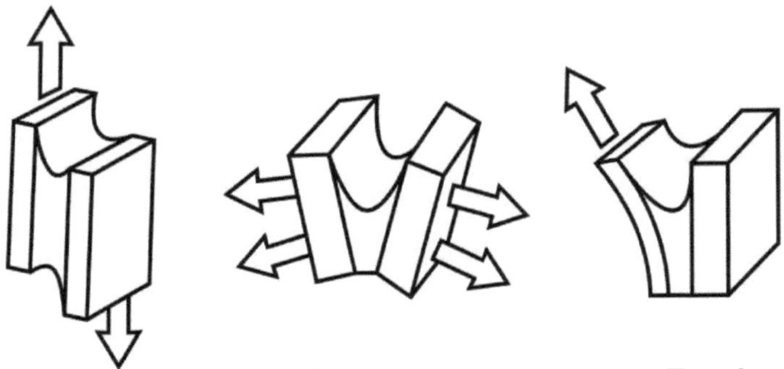

Figure 8 Three Loading Modes

To understand why warp spring-back is so damaging, we need to distinguish three ways in which a glue line can be loaded. This distinction is fundamental to everything that follows in this book.

Shear. Both bonded parts are displaced in parallel, in opposite directions, like two overlapping playing cards being pulled apart. The force distributes itself uniformly across the entire bonding area. This is the most favourable loading mode: a glue line achieves its highest load-bearing capacity under shear.

Peel. One bonded part is pulled away from the glue line at its edge, like peeling a sticker. The entire force concentrates along a linear peel front. At any given moment, only a narrow strip of the glue line is under load, and the adhesive ahead of the peel front carries no load until the front reaches it. For a PVAc dispersion on corrugated board, peel resistance is typically an order of magnitude lower than shear resistance. This is the loading mode generated by warp spring-back at the tab edges, and the primary reason warped boards produce pop-open failures.

Cleavage. The bonded parts are pulled apart from one end, like opening a book. The distinction from peel is that both substrates are rigid, so the force acts across a broader front rather than concentrating on a single line. Cleavage resistance is higher than peel resistance but substantially lower than shear resistance.

In practice, these loading modes rarely appear in isolation. A familiar example of peel stress arises when the box dimensions exceed the narrowest setting of the straight-line joggers in the delivery stacker. The joggers force the pack to a width narrower than its natural geometry allows, and the resulting lateral compression causes the panels to bow outward. This bowing lifts the glue tab from the outside edge inward, generating a peel front that propagates along the joint. Even a dimensional mismatch of one or two millimetres can be sufficient to initiate this failure on a lightweight board with low green strength.

Cleavage is the characteristic failure mode of stiff constructions. When a BC double-wall or triple-wall blank is folded at the manufacturer's joint, the residual bending stiffness at the crease line resists the fold. If the crease is too shallow or the board too rigid, the panels spring back symmetrically, opening the joint from one end like a book. This is classic cleavage: both substrates are rigid, and the opening force acts across a broad front rather than concentrating on a single peel line. Pop-opening in heavy-duty packaging is almost always a cleavage-dominated failure.

Most real-world failures, however, are mixed-mode events. A warped blank arriving at the stacker with residual twist will generate peel stress at one corner of the tab and cleavage across the opposite edge simultaneously. Jogger forces and stack weight act on the joint, along with spring-back. The glue line must therefore resist not a single, well-defined loading mode but a shifting combination of peel and cleavage under variable external constraints. This is why the problem is so difficult to solve by adjusting a single parameter: adhesive, crease depth, machine settings and board properties interact as a system, and the failure mode changes with each combination.

The practical consequence is decisive. At the manufacturer's joint, the loading mode is almost never shear. Warp spring-back pulls the tab open from the edge: that is peel. Twist introduces an asymmetric opening component that combines peel and cleavage. A uniform, area-distributed tensile load – the only mode in which the full bonding area contributes to resistance – virtually never occurs in practice. The glue line is therefore always loaded in the mode for which it is weakest, and the green strength must be sufficient to withstand precisely this unfavourable condition.

3.3 The Liner as the Unknown Variable

Not all corrugated board is equal, and the differences between liner types are dramatic for MJ bonding. Kraftliner, testliner, recycled liner (fluting/schrenz), white-top, and coated liners differ in their strike-in behaviour by a factor of 2-4. Strike-in is not a pure substrate property; it is an interaction between substrate and adhesive. On the substrate side, porosity, capillary radius distribution and surface sizing determine the absorption rate. On the adhesive side, viscosity, solids content, PVAc particle size and the surface tension of the serum phase all influence how quickly and how deeply water penetrates the substrate. Pizzi and Mittal (2017) describe this process as capillary-driven filtration [31]; Schneberger (1983) adds that on highly porous substrates, the depth of serum penetration can become so great that the adhesive film is starved at the interface, a condition known as a starved joint [32].

In practice, this means that standard PVAc dispersions are formulated for kraftliner. On that reference substrate, they offer a balanced combination of open time, green strength and film-forming rate. Problems arise when the same adhesive is used on different substrates. On recycled liner, it can lose its wetting capacity within fractions of a second; on coated liner, green strength develops so slowly that the pack springs open in the delivery stacker. The combination of liner and adhesive determines the outcome, not either factor alone.

Liner type	$Cobb_{60}$	Strike-in behaviour	Implication for adhesive selection
Kraftliner	35–45 g/m²	Moderate absorption; standard dispersions are formulated for this substrate	Reference substrate; standard adhesive is appropriate
Testliner	50–70 g/m²	Faster strike-in than kraftliner; shortened open time	Dispersion with longer open time (higher solids, retarder)
Recycled liner	80–120 g/m²	Very fast strike-in; serum penetrates deeply; risk of bleed-through	Higher viscosity and reduced application weight
Coated liner	8–18 g/m²	Strike-in strongly inhibited; green strength delayed	Specialist formulation with fine particle size; high pop-opening risk

$Cobb_{60}$ (ISO 535:2023 [75]) measures the water uptake of the liner surface over 60 seconds and is the standard laboratory proxy for liner porosity. The same Cobb value that governs adhesive strike-in at the MJ also governs the waterscoring dose required at the crease line; Chapter 11 treats this explicitly.

Particularly problematic is the asymmetric construction at the MJ: two different surfaces are frequently bonded together, the glue tab (outer liner, often kraftliner) and the opposing surface (inner liner, often testliner or recycled liner). Water migrates preferentially towards the more porous side, film formation is asymmetric, and adhesion to the denser surface can suffer as a result. Moisture further compromises the starting conditions.

Recycled liners respond disproportionately to climatic variation: at 90 per cent relative humidity, the elastic modulus of recycled paper falls by approximately 70 per cent, compared with only around 30 per cent for kraftliner[10]. Humid storage conditions, therefore, mean more warp and slower strike-in, and both effects act in the same unfavourable direction.

3.4 MJ Bonding with Hot Melt: Where the Physics Differs

The preceding sections have concentrated on PVAc dispersion adhesive because it dominates European FFG practice for standard corrugated packaging. In several important segments – heavy-duty packaging, shelf-ready displays, difficult substrates such as clay-coated or polymer-laminated liners, and much of the US market for corrugated displays and shippers – hot melt adhesive is the standard choice, and the physics of the manufacturer's joint changes substantially. The underlying chemistry (typically EVA-based ethylene-vinyl-acetate hot melts, with metallocene and APAO grades growing in share) is discussed in Pizzi and Mittal (2017)[31]; Hallbäck et al. (2014) provide a finite-element treatment specifically of hot-melt joints in carton board[74].

Three differences matter for the MJ.

No strike-in, no water transport. A hot melt is a 100 per cent solids system: it carries no water, does not wet the substrate by capillary action, and therefore depends neither on liner porosity nor on $Cobb_{60}$. The Cobb-driven selection logic of Section 3.3 largely falls away. A coated liner that is hostile to PVAc is, from the hot-melt perspective, almost indistinguishable from a kraftliner. This is why hot melt is the default choice on coated and laminated substrates.

Green strength develops in fractions of a second. Bond build-up is driven by cooling rather than by strike-in: as the molten bead crosses its solidification temperature (typically 70 to 110 °C for EVA systems, lower for APAO and reactive PUR grades), viscosity rises by several orders of magnitude within a second and the joint passes into a load-bearing state. Open time is correspondingly short, typically well below one second for standard hot melts, which is why hot-melt MJ lines are usually configured with a compression section immediately downstream of the nozzle rather than an extended belt press. Pop-opening in the delivery stacker is essentially eliminated; the failure modes migrate elsewhere.

Thermal tolerances replace moisture tolerances. What hot melt gains in substrate-independence, it loses in thermal sensitivity. Substrate temperature below approximately 15 °C allows the bead to set before full contact, producing a cold-set joint with poor wet-out; substrate temperature above approximately 40 °C delays setting into the press section and can cause bead displacement.

Adhesive temperature must be controlled to within a few degrees of the target range; a drift of 10 °C at the nozzle is enough to shift the joint from over-wet (bleed-through, stringing) to under-wet (cold-set failure). Whereas PVAc forgives temperature variation and punishes liner variation, hot melt forgives liner variation and punishes temperature variation. The failure mode inventory, in other words, is different – not smaller.

For the structural analysis that follows – loading modes (3.2), mechanisms of uneven bonding (3.5), flute-phase dependence of crease quality (3.6), narrow-tab geometry (3.7) and lightweighting (3.8) – the conclusions are largely adhesive-independent: they follow from the geometry of corrugated board and from the mechanics of spring-back, not from the chemistry of the bead. Where an effect is adhesive-specific, the text will say so. Chapters 9 and 10 return to hot melt in the context of stitching and damage-tolerant MJ design.

3.5 Three Mechanisms of Uneven Bonding at the MJ

As the preceding sections have made clear, MJ bonding is inherently asymmetric by virtue of the basic geometry of corrugated board. Washboarding sharpens this asymmetry through three interconnected mechanisms that reinforce one another. But even without measurable washboarding, the underlying tendencies are present: lightweight liners, deep flute profiles and high machine speeds are sufficient for these mechanisms to take effect.

Mechanism 1: Adhesive displacement. Even on nominally flat corrugated board, nip pressure during joining concentrates at the flute tips because the liner is supported there. In the valleys, it deflects, and local contact pressure is lower. Where washboarding is also present, and a tip-to-tip configuration faces the opposing surface, this effect is massively amplified: adhesive is actively displaced from the peaks into the valleys. This mechanism is faster and more forceful than the gravity-driven flow described in Phase 2.

Mechanism 2: Differential substrate impregnation. The flute tips have been pre-treated by the corrugator starch adhesive through flute-tip densification. The PVAc adhesive strikes in more slowly there, remains flowable for longer and is displaced more readily during joining. In the valleys, by contrast, it becomes immobile quickly, even though there is little or no contact with the opposing surface during joining.

Mechanism 3: Contact area deficit. Conde et al. (2012) confirm that the width of the effective bonding zone dominates the fracture load [21]. The basic geometry alone produces closer contact at the flute tips than in the valleys. Washboarding sharpens this deficit dramatically: the effective contact area contracts to the peaks, and it is precisely there, as Mechanisms 1 and 2 demonstrate, that the adhesive is thinnest or absent altogether.

This mechanism is not limited to the micro-scale geometry of flute tips and valleys. Warp and residual internal stresses in the board produce the same effect at a macro scale. When a warped blank enters the joining section, the two mating surfaces do not lie flat against one another.

The belt pressure must first overcome the warp before any contact occurs. On a lightweight single-wall with moderate warp, the available pressure is usually sufficient. But where the warp is pronounced, where S-warp creates opposing curvatures along the tab length, or where twist lifts one corner of the glue tab away from the opposing surface, the belt pressure may not be high enough to establish full contact across the entire bond area. Parts of the adhesive bead never touch the opposing liner.

The problem is compounded as soon as the belt pressure is released. Even when the press forces the surfaces together during joining, the elastic recovery of the warped board pulls them apart before the adhesive has developed sufficient green strength to hold them.

The contact area under pressure is not the same as the contact area that remains after the pack leaves the belt section. In this sense, warp does not merely reduce the initial contact area; it actively works against the bond during the critical transition from Phase 3 to Phase 5. This is why nominally identical adhesive quantities and machine settings can produce reliable joints on a flat board and consistent failures on a warped board from the same production run.

3.6 Crease Quality and Flute Phase Position

Creasing at the MJ has an ambivalent effect. On the one hand, it reduces bending stiffness at the fold edge: Garbowski et al. (2021) quantify a stiffness loss of 10 to 30 per cent from the creasing operation alone[4]. Cannella and Dai (2006) show that as folding progresses, stiffness can fall by a factor of 1,400, at which point the plies delaminate completely[20]. This weakening relieves the glue line by reducing the spring-back force.

On the other hand, the crease line concentrates deformation into a narrow zone directly adjacent to the glue line. The glue line sits at the transition from the stiff tab area to the softened crease zone – precisely at the point of maximum stress concentration. DIN 19304 defines the fold factor as the governing parameter [22]: insufficient creasing increases fold resistance, while excessive creasing destroys the liner and creates crack-initiation points.

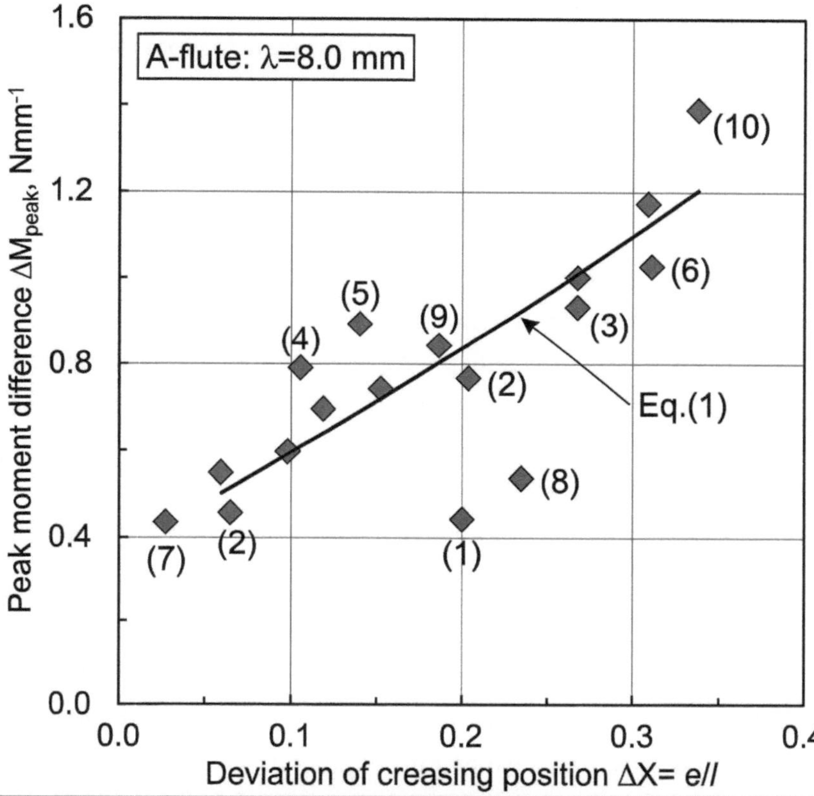

c_p/t	0.09	0.13	0.16	0.19	0.22	0.25	0.28	0.32	0.35	0.38
Case	(1)	(2)	(3)	(4)	(5)	(6)	(7)	(8)	(9)	(10)

Figure 9 Influence of crease position relative to the flute geometry on the peak spring-back moment for A-flute corrugated board (λ = 8.0 mm). The horizontal axis shows the dimensionless creasing-position deviation ΔX = e/λ, where e is the distance of the creasing tool from the flute valley and λ the flute pitch. The vertical axis shows the peak moment difference ΔM_peak (N·mm^{-1}) measured at the crease during the first fold. Each data point (1)–(10) corresponds to a different ratio of creasing-penetration depth to board thickness c_p/t, increasing from 0.09 (case 1) to 0.38 (case 10), as listed in the table beneath the chart. The solid line represents the empirical fit Eq. (1), ΔM_peak = 2.21 · (e/λ) + 0.34. Source: Nagasawa & Komiyama (2012), endnote [29].

Nagasawa and Komiyama (2012)[29] quantified the influence of a frequently overlooked factor in practice: the position of the creasing tool relative to the flute geometry. When the creasing tool strikes a flute tip (apex), the frequency of liner fracture is at its maximum. On the flute shoulder, an asymmetric force introduction results. In the valley, fracture frequency is lowest, but the risk of excessive crease depth increases.

The plot in the above figure makes the central physical claim of Section 6.3 quantitative. As the crease drifts away from the flute valley (ΔX rising from 0 toward 0.4—up to 40% of a flute pitch), the peak moment that the fresh fold must overcome rises from roughly 0.5 $N{\cdot}mm^{-1}$ to about 1.2 $N{\cdot}mm^{-1}$. The slope of the fit is 2.21 $N{\cdot}mm^{-1}$ per unit of e/λ. This is the quantitative heart of the relationship. A creasing position halfway between valley and tip ($\Delta X \approx 0.25$) raises the peak moment by about a factor of two over the valley case.

The vertical scatter at any given ΔX is a second-order effect. For example, case (8) at 0.53 versus case (10) at 1.40, with almost the same ΔX, differs due to the penetration depth ratio c_p/t. Deeper penetration partly relaxes the position dependence by softening the substrate before the fold begins. Shallow penetration leaves the fold more exposed to positional effects.

Two practical readings follow.

First, the variable that drives the spring-back force at the crease is geometric, not material. It depends on where the tool lands on the flute wave.

Second, that position varies inherently across the web (Section 7.1, cut register and skew). As a result, the spring-back moment scatters at the joint by a factor of 2 even when all machine settings are correct. This makes the case for waterscoring as a local-conditioning intervention rather than a global parameter adjustment.

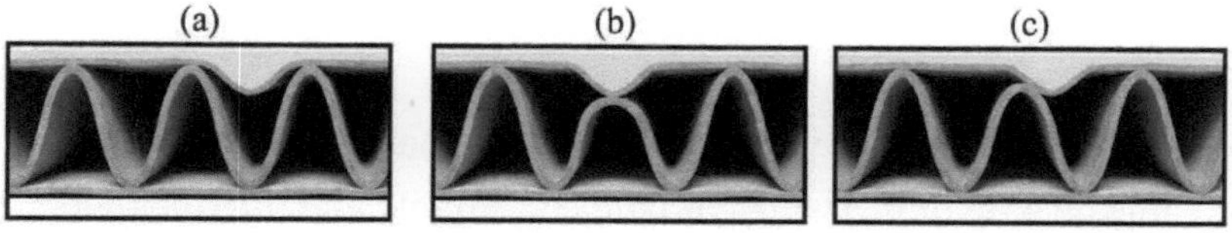

Figure 10 The three contact cases between the creasing tool and the flute geometry, shown in cross section for two flute pitches (top row, larger pitch; bottom row, smaller pitch). (a) Tool over a flute valley: the rule presses into the trough between two adjacent flute peaks, and the indentation in the top liner sits in the unsupported span between flutes. (b) Tool over a flute tip: the rule lands directly on a flute peak, which carries the load to the bottom liner along the shortest possible path. (c) Tool over a flute flank: the rule meets the inclined wall of a flute, and the indentation drifts toward the side of least resistance. Source: Nica et al. (2020), endnote [30].

Nica et al. (2020)[30] confirmed this finding for inline creasing directly at the corrugator. Because the crease is applied in the machine direction, the phase relationship between the creasing tool and the flute geometry varies both across the web width and over time. The three panels visualise why the flute phase position is treated in this book as the dominant geometric variation driver. Each contact case loads the same paper composite in a fundamentally different way.

In case (b), the tool, the flute peak and the bottom liner form a short, stiff column. The board behaves almost like a solid panel: the indentation is shallow, the local crushing is concentrated on the flute tip, and the spring-back moment after the fold is high. This is the geometry that Nagasawa and Komiyama parameterise as the upper bound of the e/λ relationship shown in Figure 7.X (the preceding figure).

In case (a), the tool sits over a span. There is no rigid column beneath it; the top liner deflects further before the rule transfers force to the next flute pair, the local crushing spreads laterally, and the spring-back moment after the fold is the lowest of the three. This is the lower bound of the same relationship.

Case (c) is the practical reality on a running line. The crease almost never lands cleanly on a tip or in a valley; it lands on a flank, and the rule then follows the path of least resistance. The flute tips sideways, the contact point drifts in the direction of the weaker flank, and the indentation acquires a lateral component the operator has not asked for. This is the mechanism behind the meandering crease line that produces fishtailing, and it is the reason that even a perfectly aligned creasing rule produces visibly different fold quality from one blank to the next when the upstream cut register has drifted.

The two rows make a further point. The same three cases recur on every flute pitch, but the dimensional consequences scale with the pitch. On A-flute ($\lambda \approx 8$ mm), a millimetre of position drift moves the rule by an eighth of a flute; on E-flute ($\lambda \approx 3$ mm), the same millimetre moves it by a third.

The result is an unavoidable scatter in the point of contact, tip, shoulder or valley, and therefore an unavoidable scatter in crease quality. This insight becomes central in Chapter 11, on waterscoring.

3.7 Narrow Glue Tabs: Less Area, More Risk

The European Packaging Regulation PPWR (EU 2025/40)[26] and economic pressure on material consumption are driving tab widths downward: from the typical 35 to 40 mm range to 20 to 25 mm, and in extreme cases, below 20 mm. What appears at first glance to be a simple geometric change fundamentally alters the mechanics of loading.

A standard 35 mm tab accommodates three adhesive beads with adequate spacing between them. At 22 mm, two beads are the maximum; below 18 mm, only one is possible. A single bead means that if it lands on a washboarding peak, on a pre-impregnated zone or in an area of contact loss caused by warp, the entire bond can fail. Every redundancy is gone.

Bending Stiffness and Peel Stress

Bending stiffness scales approximately with the cube of the flute height. A single-wall E-flute board has a stiffness of approximately 2,500 to 3,500 N·mm; a BC double-wall reaches 38,000 to 40,700 N·mm; a triple-wall board 80,000 to 250,000 N·mm. Taking into account the effect of creasing, which removes 40 to 50 per cent of stiffness in single-wall but only 20 to 30 per cent in triple-wall because the inner plies are barely reached, the following picture emerges.

Scenario	Uncreased stiffness	After creasing	Peel stress (normalised)
B-flute, 35 mm	≈ 5,200 N·mm	≈ 50 % → 2,600	1.0× (reference)
B-flute, 22 mm	≈ 5,200 N·mm	≈ 50 % → 2,600	≈ 1.6×
C-flute, 35 mm	≈ 10,500 N·mm	≈ 55 % → 5,800	≈ 2.2×
C-flute, 22 mm	≈ 10,500 N·mm	≈ 55 % → 5,800	≈ 3.5×
BC double-wall, 35 mm	≈ 38,000 N·mm	≈ 60 % → 22,800	≈ 8.8×
BC double-wall, 22 mm	≈ 38,000 N·mm	≈ 60 % → 22,800	≈ 14×
Triple-wall, 22 mm	≈ 100,000–180,000 N·mm	≈ 70–80 %	≈ 25–50×

Derivation of the peel-stress ratio. The peel stress generated at the tab edge by the spring-back of a folded panel is, to a first approximation, proportional to the bending moment per unit tab length, and therefore to the creased bending stiffness EI^c divided by the tab width w. The values in the last column are normalised to the reference case (B-flute, 35 mm creased): $\sigma_{peel} / \sigma_{ref} = (EI^c / w) / (EI^{c},_{ref} / w_{ref})$. The cube-of-flute-height scaling enters indirectly through EI^c; it is not applied a second time to the tab width. The values assume pure peel at the tab edge; the real loading is a mix of peel and cleavage, with the ratios varying but the rank order preserved.

The figures are striking. A BC double-wall board with a 22 mm tab generates fourteen times the peel stress of the reference case. For triple-wall, the factor ranges from 25 to 50. These are not theoretical values; they are the loads the glue line must withstand in the first seconds after joining, while it is still dependent on green strength and film formation has barely begun.

3.8 Lighter Paper Weights: Less Material, Narrower Tolerances

The industry trend towards lighter paper weights (downgauging) simultaneously worsens the starting conditions on several levels. Lighter liners exhibit more pronounced washboarding because less fibre mass is available to compensate for the localised deformation at the flute tip bonding points. They are more sensitive to moisture because less fibre mass acts as a buffer. Their strike-in is faster, but the substrate's absorption capacity can be exhausted before sufficient water is removed, leaving the paper saturated before film formation is complete.

The tolerance windows for adhesive quantity, viscosity and process control narrow accordingly. What worked with a 200 g/m² kraftliner fails with a 120

g/m^2 testliner. The combination of a narrower tab, lighter paper and higher board stiffness – because the customer expects the same performance from less material – creates a triangle of conflicting demands that conventional setting optimisation can barely manage.

3.9 The Complete Picture: Why the MJ Is the Weak Point

From the flute geometry. The basic construction of corrugated board already produces uneven contact conditions. The liner deflects in the valleys; nip pressure concentrates at the tips. This fundamental effect exists independently of any defect and becomes more pronounced the lighter the liner.

From the corrugator. Washboarding (modulated surface, differential porosity), warp (stored spring-back energy), and possibly crushing damage, all of them amplifiers of the inherent geometric asymmetry.

From the liner type. Strike-in behaviour varying by a factor of two to four, which determines film formation rate and open time – for dispersion adhesives; hot melt shifts the sensitivity from porosity to temperature, as Section 3.4 sets out.

From the geometry. Progressively narrower tabs that increase peel stress disproportionately and eliminate redundancy.

From the creasing. Variable contact position of the creasing tool on the flute structure, producing an unavoidable scatter in crease quality.

From the loading mode. Peel and cleavage, rather than shear, are the least favourable loading modes acting on a glue line that is only seconds old.

The MJ is not the weak point because it is poorly designed. It is the weak point because the basic geometry of corrugated board does not permit homogeneous bonding, and because every individual defect generated in the preceding process steps converges at this single point of failure. What was an acceptable compromise at the corrugator accumulates at the folder-gluer into a problem that conventional parameter optimisation can no longer contain.

Chapter 4:
Dimensional Deviation as a Multiplier:
From the Line to the Retail Shelf

Introduction card

Dimensional deviation is not only a converting defect. It is a cost multiplier. This chapter follows a weakened or off-dimension box through the case erector, filling line, pallet, transport, loading bay, and retail shelf. It is relevant for readers who need to locate where losses arise, how they compound, and why they are often misclassified in OEE systems.

"*How does a millimetre at the folder-gluer become a million euros at retail, and where exactly along the chain does the cost compound?*"

Economic impact: The cost cascade has several measurable nodes. At the case erector, poor crease quality can cause 10 to 20 micro-stops per shift, each lasting 30 to 180 seconds. This can accumulate to 15 to 30 minutes of downtime per shift. At an FMCG line cost in the EUR 3 000 to 40 000 per hour range, the upper bracket anchored by the PMMI 2023 figure of approximately EUR 25 000 per hour for high-volume beverage and household-care lines, the lower bracket characterising mid-volume FMCG and low-volume specialty packing, one affected shift can generate EUR 1 000 to 12 000 in directly attributable cost depending on the line class. If the case erector stops, the filler and labeller often stop as well. At pallet level, non-parallel stacking weakens load transfer and can reduce service BCT by 30 to 50 per cent. At retail, the consequences include returns, complaints, chargebacks, supplier-rating penalties, and lost repeat purchase.

Operational efficiency: Crease-quality-related disruptions often account for 15 to 20 per cent of total production-time loss on affected lines. These losses are usually hidden in OEE systems under broad categories such as "material defects" or "miscellaneous downtime". Separating crease quality from these generic buckets is necessary before the cause can be corrected. A tolerance of 1.5 mm creates measurably more stoppages than a tolerance of 1.0 mm on the same machine.

Ecological impact: Each rejected or returned pack carries the carbon footprint of the corrugated board and the usually much larger footprint of the product inside it. For food packaging, the product-to-packaging carbon ratio can range from 10:1 to 100:1. The dominant ecological cost of a damaged FMCG pack is therefore often the wasted product, not the wasted box.

Quality: This chapter maps the main failure modes from line to shelf: pop-opens, jams, gap deviation, stack instability, transport damage, pallet rejection, and retail returns. It helps readers locate where the loss becomes visible and where the original cause may sit.

Regulatory context: PPWR supports precision rather than simple material substitution. A process that prevents specification failures improves the failure side of PPWR compliance without necessarily changing the substrate.

Limitations: The economic figures are order-of-magnitude values. Pack-specific accuracy requires line-by-line analysis. The EUR 25,000 per hour figure is an FMCG anchor; lower-volume speciality packing can be several times lower.

Read in full: line managers, quality managers, retail supply-chain owners.

Card sufficient: executives building the business case for line-level intervention.

Prerequisite: Chapter 3.

Follow-up: Chapter 5, the ecological cascade.

In the preceding chapters, we have seen how defects originate at the corrugator and how the manufacturer's joint condenses them into a single point of failure. We now leave the factory and follow a box with a weakened MJ as it travels through the value chain. The conclusion is sobering: what begins as a millimetre-scale deviation multiplies at every stage.

The medical analogy remains useful here. In Part I, we described the anatomy (Chapter 1) and the pathology (Chapters 2 and 3). What follows now is the symptomatology: the visible and measurable consequences of the condition, stage by stage.

A note on scope. This chapter follows the dominant loss pathway in fast-moving consumer goods (FMCG): case erectors, mixed pallets, retail loading bays, and e-commerce returns. Heavy-duty industrial palletising – automotive parts, chemicals, building materials, tier-one B2B supply – runs on a different failure signature: fewer SKUs, higher per-unit value, denser and often sharper contents, bespoke testing regimes, and a cost structure in which a single rejected load can exceed an entire FMCG shift's chargeback exposure. The multiplier principle is the same in both worlds; the mechanisms, the numbers and the instruments of recovery are not. Where the distinction matters below, it is flagged explicitly.

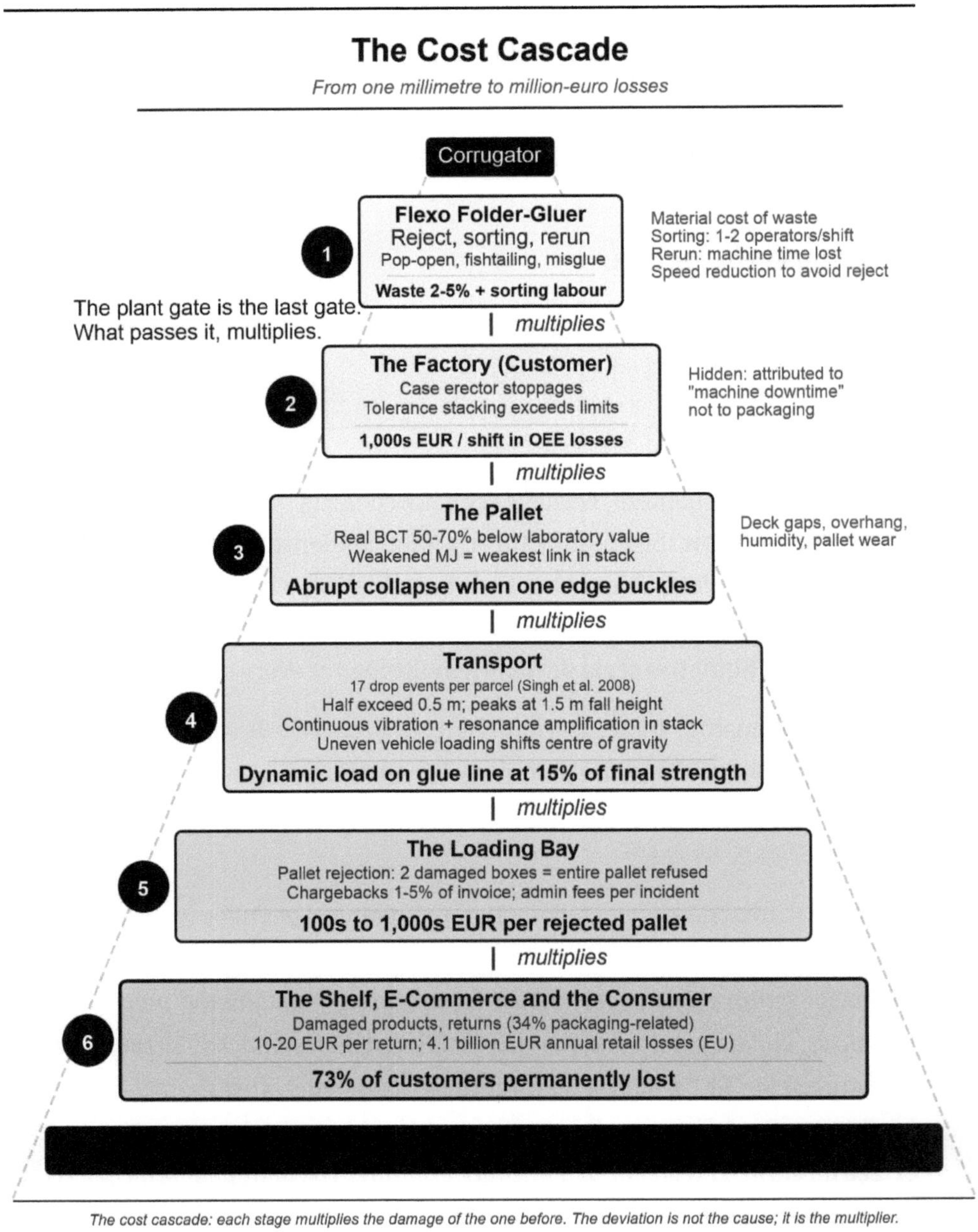

Figure 11 The cost cascade

4.1 Disruptions in Linked Production Systems

The first point at which an off-dimension box causes damage is the customer's automated packing line. Case erectors are designed to erect, fold and close flat blanks in fractions of a second. They operate within tight tolerances: ±3 mm for die-cut boxes and ±5 mm for RSC boxes. For the manufacturer's joint, the permitted fishtailing tolerance is 4.5 mm [1, 2].

Fishtailing and Gap Deviations

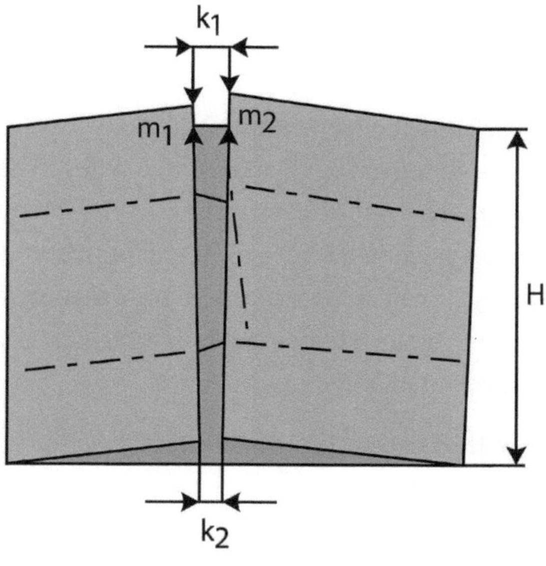

Figure 12 Geometry of the manufacturer's joint as measured per FEFCO convention. The view is onto the outside of the erected box at the seam where the two end flaps meet. H is the box height (the length of the joint). k_1 and k_2 are the gap widths at the top and bottom of the joint respectively, measured between the two facing edges of the meeting flaps. m_1 and m_2 are the approach widths of the two flaps at the top edge — i.e., the visible width of each flap immediately adjacent to the gap. Dashed lines indicate the horizontal score lines on the side panels. Fishtailing is defined as the absolute

difference $|k_1 - k_2|$; a non-zero difference identifies a tapered joint, the case-style tolerance for which is set by the FEFCO standard.

Fishtailing describes the lack of parallelism at the manufacturer's joint: one side of the glue line sits higher than the other, giving the box a fishtail appearance when viewed from above. The primary cause lies in the crease itself: when the crease line follows a meandering rather than a straight path, an uneven fold angle develops as the side panels are erected, resulting in a fishtail on the finished box. A common trigger for meandering crease lines is the crossing of the flute lay.

Where the creasing tool meets the flute at an angle rather than perpendicularly, the crease line deflects laterally because the resistance of the flute structure is direction-dependent. With every transition from tip to valley, the lateral resistance changes, and the tool follows the path of least resistance rather than the prescribed straight line. The interaction between warp spring-back and asymmetric MJ bonding reinforces the effect further: where the glue line is weaker on one side, because washboarding has displaced the adhesive there, that side yields more readily under the spring-back force than the other, amplifying the fishtail.

The consequence in the case erector is immediate: a fishtailing box sits skewed in the magazine, is picked up off-centre and jams during erection. The operator must intervene, and the machine stops.

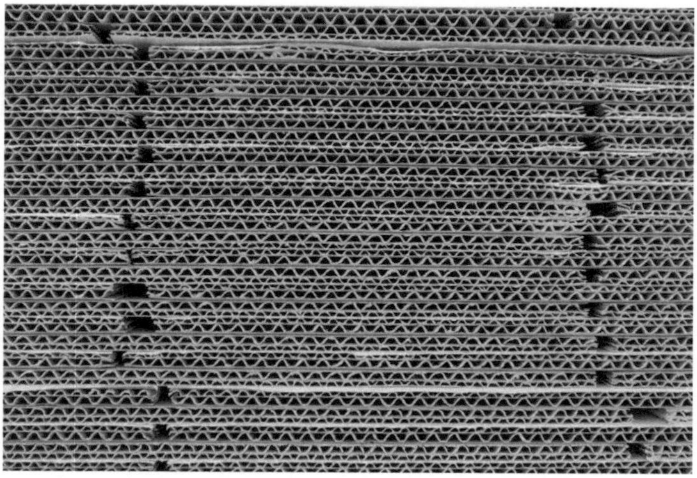

Figure 13 Gap deviations are most easily detected in the bundle of knock-down flats (KDFs). Viewed end-on, a strapped bundle of glued, flat-folded boxes presents the manufacturer's joint of every flat in a single stacked column. Variations in gap width, parallelism and crease quality that are difficult to identify on an individual flat become legible at a glance across the bundle: a uniform offset reveals itself as a column of identically displaced gaps, a tapered (fishtail) gap appears as a column that fans open from one edge to the other, and an erratic gap stands out as a discontinuity against its neighbours. The bundle thus serves as a no-cost inspection sheet, written in the product's own language — provided the shift leader is trained to read it.

Fishtailing is not the only geometric deviation that matters at the manufacturer's joint. A second, independent failure mode is the parallel shift of the gap itself: the two edges of the glue line remain parallel to each other, but the overlap lies systematically outside the specified value across the full height of the box. A common cause is the varying flute position at the crease line: when the creasing tool strikes a flute apex, the crease line deflects differently than when it strikes a valley, and the effective fold position shifts by up to half a flute pitch. Secondary fracture lines adjacent to the primary crease can also displace the actual fold point and, with it, the gap dimension.

The consequences are critical in both directions. Where the gap is too tight, the box may be impossible to erect, or the internal dimensions fall below the specified value, and the product no longer fits. Where the gap is too wide, the box is oversized and loses stacking strength.

Cascade Effects in Linked Production Systems

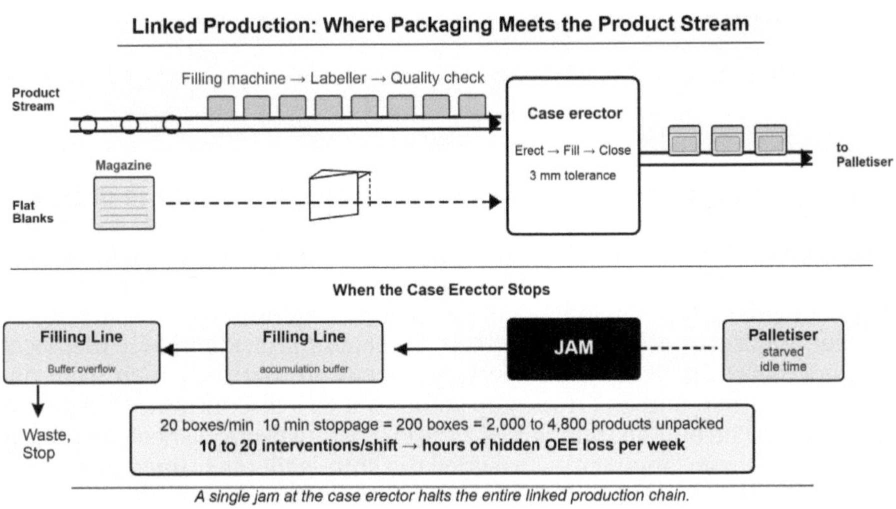

Figure 14 The case erector as the bottleneck of linked packaging production. Top panel: the normal operating state. The product stream (filling, labelling, quality check) and the flat-blank stream from the magazine converge at the case erector, which performs the three sequential operations Erect → Fill → Close within a 3 mm dimensional tolerance window before passing the filled cases to the palletiser. Bottom panel: the consequence of a single case-erector jam. Upstream, the filling line first absorbs the stop in its accumulation buffer, then runs the buffer to overflow, then halts (waste and line stop). Downstream, the palletiser is starved and accrues idle time. At a nominal rate of twenty boxes per minute, a ten-minute jam translates into 200 unfilled cases and, depending on count per case, 2,000 to 4,800 products that fail to leave the line. With ten to twenty such interventions per shift — a typical figure for plants running on tight blank tolerances — the cumulative OEE loss reaches several hours per week, most of which is invisible in headline

production reports because no individual stoppage is long enough to register as a downtime event.

Modern packing lines are linked. The filling machine, labeller, quality control, case erector, and palletiser form a chain, with each station relying on the previous one. If the case erector stops for thirty seconds, the entire upstream line backs up. Many installations lack accumulation buffers due to space and cost constraints, so the filling line stops immediately when it is blocked. The case erector holds a key position. It is the only station where two streams, products from filling and packaging from the blank magazine, must arrive together. Accumulation buffers can absorb temporary mismatches, for example, between filling and labelling or palletising and shipping. However, there is no buffer for the case erector, as the product cannot wait inside an unerected box. The case erector's dimensional tolerance, typically 3 mm, reduces the line's timing tolerance to zero just as millimetre-scale variations discussed in Chapters 6 and 7 reach the customer.

The costs of such stoppages are considerable, yet rarely match initial estimates. Typical packing machines for corrugated shippers run at 15-25 boxes per minute. Each box contains multiple products, such as ten cake trays or 24 beverage cans. If the line runs at 20 boxes per minute, a ten-minute stoppage means missing 200 shippers or 2,000 to 4,800 individual products. Whether those units are recovered in the same shift, deferred, or lost entirely depends on line capacity and headroom. Direct costs are easier to price. An idle FMCG packing line incurs hundreds, or even thousands, of euros per hour, covering labour, overheads, depreciation, and knock-on scheduling costs. When operators intervene 10 to 20 times per shift, hidden OEE losses add up to hours of lost production each week. The implication for procurement is uncomfortable. A converter quoting blank price per thousand without jam rate per thousand quotes the wrong unit, since the plant pays back the price difference with interest in OEE.

These losses are especially insidious because they are rarely linked to packaging material. Production statistics list them as machine downtime or operator error, not packaging defect. The true cause, a millimetre-scale problem at the MJ, stays hidden. The case erector jam is the visible end of a causal chain that began at the corrugator the night before. A drift in the cut register, a moisture gradient in the pallet, or a residual crushing pattern from a worn nip may all be clear on bundles long before they reach the loading bay (Chapter 7).

Besides dimension-related defects, there is a second common failure: adhesive application defects at the folder-gluer. Adhesive splatter, satellite deposits, and misplaced glue beads can cause two-pack panels to bond in isolated spots, known as inside gluers or outside gluers. For the case erector, these unintended bonds matter. They can block the erection or damage the liner when panels are pulled apart. Chapter 9 addresses, in detail, mechanisms and remedies for adhesive application quality.

Tolerance Stacking at the Case Erector

Source of Error	Tolerance	Cumulative	Assessment
RSC base dimension	±5.0 mm	±5.0 mm	Specification
Fishtailing (MJ)	up to 4.5 mm	±9.5 mm	Critical
Warp spring-back	2 to 3 mm	±12.5 mm	Borderline
Crease line scatter	±1.0 mm	±13.5 mm	Exceeded

The table illustrates how individual tolerances accumulate. Each error in isolation may lie within specification; together, they breach the tolerance band of the case erector. This is precisely the fundamental problem: it is not the individual defect but the sum of the deviations that causes failure.

The cumulative values above follow worst-case tolerancing, which assumes that all contributing errors reach their maximum simultaneously and in the same direction. A statistical (root-sum-square) combination would yield a narrower predicted band; the worst-case view is chosen here because case-erector stoppages are triggered by the unfavourable coincidence of deviations, not by their average. A line that tolerates the RSS band well will still stop when the worst-case stack occurs.

4.2 Failure on the Pallet

Figure 15 Unit-load compression test of a palletised stack of corrugated shippers. A representative sample unit load — here six identical RSC boxes arranged in a two-by-three pattern on a four-way wooden pallet — is placed between the platens of a hydraulic compression tester. The upper platen advances at a controlled crosshead speed (typically 12.7 mm·min^{-1} per ASTM D642 / ISO 12048), and the load is recorded continuously until the unit load yields. The test measures the load-bearing capacity of the system as a whole, not of an individual box: the result is jointly determined by the BCT of the boxes, the stacking configuration, the pallet deck-board geometry, the box-to-deck overhang and any unit-load stabilisation (stretch wrap, edgeboard, strapping). Source: image courtesy of the Virginia Tech Center for Unit Load Design (CULD); test methodology per DIN 55440 / ISTA 3K, with the unit-load multipliers developed by Baker et al. (2016) [38] and Quesenberry (2020) [46].

The first number is the Box Compression Test (BCT) value of the individual box. The BCT is the measure provided by the converter in the technical datasheet and is the value that the McKee formula (a standard industry estimation method introduced in 1963) uses to estimate the geometry and edge crush resistance of the corrugated structure. This number is determined by compressing a single empty box between flat steel platens (rigid plates) in the lab, with no pallet underneath and no adjacent boxes touching it. The BCT provides a clean, repeatable laboratory metric for stacking strength.

The second number is the unit load's load-bearing capacity—the actual pallet of filled boxes on a warehouse rack. This value determines whether goods arrive intact at the retailer. It is consistently lower than the sum of the individual Box Compression Tests (BCTs), often by a substantial margin. The reasons are seen in the photograph rather than in the box: a wooden pallet with deck boards that span gaps and lift parts of the bottom box off the support; overhang, where the box footprint exceeds the pallet outline; an interlocked stacking pattern, which loads corners differently from column stacking; and a stretch-wrap, whose tension either stabilises or pre-loads the stack.

Three findings from the cited unit-load literature anchor the discussion in Chapter 4. First, the stiffness of pallet deck boards (the boards forming the pallet's upper surface) alone can reduce unit-load compression strength (the maximum load the stacked unit can bear before failing) by 30 to 50 per cent relative to flat-platen BCT (Box Compression Test, a standard lab method). Thinner deck boards cause the largest reductions (Baker et al. 2016 [38]; Quesenberry 2020 [46]). Second, pallet overhang of just 25 mm (when the pallet extends beyond the load or rack support) reduces the supportable load by another 10 to 30 per cent, depending on stack height (Kim et al. 2023 [43]). Third, stacking configuration—column (boxes aligned directly above each other) versus interlocked (boxes offset between layers)—changes the load distribution between boxes by a factor that, on tall stacks, dominates the choice of corrugated grade (type of cardboard material) (Kmita-Fudalej et al. 2025 [44]).

The practical reading motivates the book's wider argument. A converter specifying the box on BCT alone, and a customer accepting that spec, agree on a number true in the lab but only loosely tied to pallet performance. The McKee value sets the upper bound. The unit-load test in the photograph sets the bound that matters.

Even if the box passes through the case erector unscathed, the next test awaits on the pallet. Here, laboratory conditions diverge from reality. The box compression test (BCT) measures stacking strength under ideal conditions: flat, rigid platens, a controlled climate, and a uniform load. Real pallets look quite different.

Reduction Factors: The Gap Between Laboratory and Practice

Frank (2014) [40] conducted a comprehensive review of the literature on stacking strength and shows that the real load-bearing capacity of a box on a pallet typically lies 50 to 70 per cent below the laboratory value. The reasons are numerous and cumulative.

Influencing Factor	Reduction	Source	Comment
Pallet deck board stiffness	15 to 37 %	Quesenberry 2020 [46]	Depends on board thickness
Deck board gaps (Euro pallet)	up to 26 %	Baker et al. 2016 [38]	More pronounced for small boxes
Overhang 25 mm	13 to 24 %	Kim et al. 2023 [43]	One side
Overhang 50 mm	29 to 40 %	Kim et al. 2023 [43]	Two sides
Interlocked stacking pattern	35 to 40 %	Kmita-Fudalej et al. 2025 [44]	vs. columnar stacking
Humidity above 60 % RH	20 to 40 %	Modzelewska 2006 [10]	Recycled liner more affected
Cumulative (worst case)	50 to 70 %	Frank 2014 [40]	Stack of individual effects

The reduction factors in the table above are drawn from studies that isolate each effect under its own test geometry: Baker et al. address deck-board gaps in small boxes, Kim et al. sweep overhang depth on single boxes, Kmita-Fudalej et al. compare columnar and interlocked stacking patterns, and Modzelewska loads a single box under controlled humidity. Their figures, therefore, cannot be added algebraically. Frank's 50 to 70 per cent range is a reasonable envelope for a realistic combination of effects, not a sum of individual worst cases. The table is intended as a menu of plausible reductions, not an additive formula.

Quesenberry (2020) [46] systematically investigated the influence of deck board stiffness. Baker et al. (2016) [38] quantify the effect of deck board gaps in the Euro pallet. Kim et al. (2023) [43] show that an overhang of just 25 mm reduces BCT by 13 to 24 per cent; at 50 mm, the reduction reaches 29 to 40 per cent.

The Three Pallet Effects

Three characteristics of the real pallet produce loading patterns that do not occur in BCT laboratory testing.

Deck board deflection. Under load, the deck boards deflect between the blocks. The box is not loaded uniformly across its base; stress concentrates at the block positions. The corners, where load-bearing capacity is highest, may sit over a gap or over a board, depending on position. This is position-dependent and cannot be known in advance.

Surface topography. Reusable pallets show signs of wear after only a few cycles, including protruding nails, splinters, and broken wood fibres. These generate discrete point loads that can locally exceed the nominal surface pressure by several multiples. The McKee formula [45], which was derived on the assumption of uniform load distribution, cannot capture this type of loading.

Deck board gaps. The Euro pallet has gaps of 40-45 mm between its deck boards. Where a box edge sits over such a gap, support is absent precisely at the point of vertical load transfer. Baker et al. (2016) [38] show that this effect is particularly pronounced for small boxes.

Moisture as an amplifier. Modzelewska (2006) [10] shows that ambient humidity above 60 per cent RH reduces BCT by 20-40 per cent, with recycled liners disproportionately affected. The combination of pallet effects and high humidity, typical of unprotected storage areas and loading bays, can reduce the actual load-bearing capacity to 30-40 per cent of the laboratory value.

Dimensional Deviation on the Pallet: When Edges and Corners Protrude

The pallet effects described above assume geometrically correct boxes. In practice, fishtailing and variations in gap dimensions at the manufacturer's joint result in boxes whose edges and corners do not align with those of the layer above or below. When a corner protrudes beyond the footprint of the stack, it becomes the sole bearing point for the entire column of boxes above it.

The full weight of the stack is concentrated on one or two edges rather than distributed across four. These overloaded edges buckle first, and only then does the remaining structure absorb the load. By then, the box geometry is already compromised, and the residual load-bearing capacity is a fraction of the original value. The mechanisms behind fishtailing and gap deviation are discussed in detail in Section 4.1; the interaction between warp, spring-back and the MJ bond is examined in Chapter 3.

The effect compounds from bottom to top. The bottom box in the stack carries the accumulated weight of all the boxes above it. If that bottom box has a protruding corner from a fish-tailed MJ, it bears the highest load at the weakest point. One buckled corner in the bottom layer can trigger a progressive collapse that works its way upward through the entire pallet, toppling what was, moments before, a stable stack.

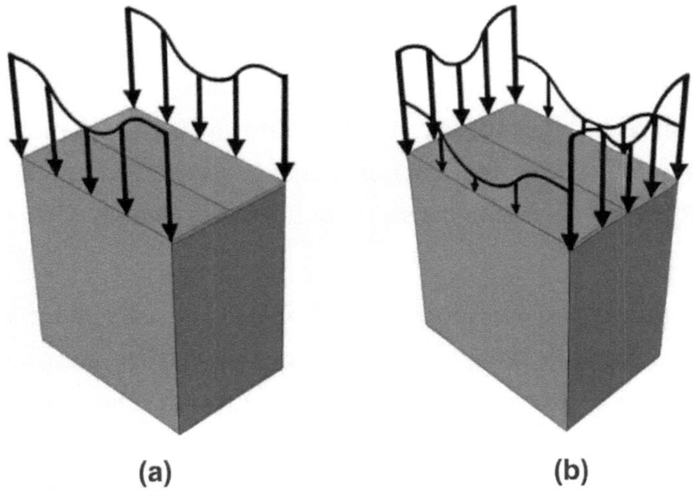

Figure 16 Vertical load distribution along the top edge of a corrugated box: (a) idealised, near-uniform pressure as assumed by the classical McKee model; (b) realistic post-buckling distribution, in which the side panels deflect outwards and the load migrates towards the four vertical corner edges, producing pronounced peaks above the corners and reduced contact pressure between them. The envelope curves above each box represent the local pressure profile. The redistribution explains why the corners, and any geometric irregularity that causes one of them to protrude beyond the stack footprint, carry a disproportionate share of the column load and therefore govern the box compression strength. (*Source: Garbowski, T. et al. [102]*

The MJ as the Weakest Link

What does this mean for a box with a weakened MJ? The stacking strength of a corrugated box depends critically on the integrity of its four vertical edges. Each vertical edge transfers the load into the layer below. When the MJ yields under load because the glue line is weaker than it should be, the box loses not just one edge but its entire load-bearing concept. Collapse does not follow proportionally; it is abrupt. As soon as one edge buckles, the remaining three edges assume the load and fail shortly afterwards.

4.3 Transport: The Forgotten Stress Phase

Figure 17 Transport as a stress phase. Once the unit load leaves the loading bay, the static stack load measured in the laboratory is overlaid with dynamic excitation: road-induced vibration, braking and acceleration peaks, lateral loads on bends, and climatic transients between heated cabin, unheated trailer and ambient road conditions. Singh et al. (2008) [48] documented drop and impact events on the order of 0.4–0.8 g during normal European parcel transport; ISTA 3K [42] formalises the corresponding test profile for FMCG supply chains. The shipper has to survive not the laboratory load, but the laboratory load plus the dynamic spectrum of the route.

Singh et al. (2008) [48] measured the real stresses in parcel shipping and established the reference figure that continues to define the discussion: on average, a parcel experiences 17 drop events on its journey from sender to recipient. Half of these drops exceed a fall height of 0.5 metres; individual drops reach 1.5 metres. Each impact loads the box dynamically, and a glue line that holds on the packing line at 15 per cent of its final strength can fail at the fifth or tenth drop.

Drops, however, are only the most visible stress. Between them lies a continuous barrage of lower-magnitude forces that cumulatively can be just as damaging.

Vibration. Every truck transmits a spectrum of vibrations through its load bed, driven by engine frequency, road surface and vehicle speed. At certain frequencies, the stack enters resonance: the boxes amplify the vibration rather than absorbing it. This cyclic loading is particularly dangerous for a freshly glued manufacturer's joint, where the adhesive has not yet reached full film strength – a state discussed in detail in Chapter 3, Phase 3 – and the bond fatigues under repetitive strain.

Figure 18 Uneven loading in the last-mile leg. Once a shipment leaves the structured environment of pallet handling and enters single-parcel distribution, the controlled stacking patterns of the warehouse are replaced by ad-hoc loading: cartons of differing footprints, weights and stiffness are placed on top of one another in the order in which they are picked rather than in a load-bearing sequence. The load distribution inside the van bears no relation to the stacking pattern the converter assumed at the design stage. Heavy parcels routinely sit on lighter ones, narrow footprints rest on overhanging edges, and "Fragile" markings function more as an after-the-fact disclaimer than as a sorting instruction. Singh et al. (2008) [48] measured drop heights of 0.5 to 1.2 m and impact accelerations beyond 0.8 g as routine events on this leg of the chain. The corrugated shipper that performs adequately under the steady, axially aligned load of a palletised stack faces, in the back of the delivery van, a fundamentally different and far less forgiving load case.

Uneven loading. Mixed pallets rarely hold boxes of the same size. Large shippers sit beside small, and heavy trays next to light displays. This causes uneven weight, shifting the centre of gravity off the pallet. During turns or braking, this generates lateral forces that stress the weakest stack joints.

Climate fluctuations. A truck crossing the Alps or standing overnight in a loading bay is exposed to temperature swings of 20 degrees or more within hours. The humidity follows: warm air holds more moisture, and as the temperature drops, the relative humidity inside the trailer climbs steeply. Corrugated board absorbs this moisture on timescales of minutes to hours; the liner swells, the fluting softens, and the box loses stacking strength precisely when the vibration loads are highest.

Insufficient load securing. When pallets are not adequately strapped or stretch-wrapped, the individual boxes must absorb the full inertia of every acceleration and deceleration event. Even a routine motorway lane change can generate 0.5 g of lateral acceleration, enough to shear the top layer of boxes sideways by several centimetres.

Throw-loading in e-commerce. In parcel logistics, sorting hubs at the last mile handle thousands of parcels per hour. The transition from conveyor to chute and manual throw-loading exposes each parcel to repeated impacts from multiple directions. Unlike a palletised stack, where boxes rest in a fixed orientation, parcels in a hub tumble freely, making every face a potential impact surface. As a result, the manufacturer's joint is frequently loaded in directions for which it was never designed.

These transport stresses form a sustained assault that laboratory tests cannot replicate. Unlike the BCT's single static load, real transport delivers thousands of minor loads that cumulatively devastate marginal bonds.

4.4 How the Retail Trade Responds to Failure

The retail trade's response to packaging failure is unambiguous: it passes the costs back to the supplier. The instruments for doing so are chargeback programmes, pallet rejection and increasingly stringent testing requirements.

Chargeback Programmes and Supplier Assessment

Amazon operates its SIPP programme (Ships in Product Packaging) across its European marketplaces (DE, FR, IT, ES) [37]. Suppliers whose packaging is robust enough to ship without an outer carton receive incentives of up to four euros per unit, depending on product category; those who fail to meet the requirements pay higher fulfilment fees and repackaging costs. The principle in Europe is the same as in the United States: the packaging must survive the shipment without an outer box.

In bricks-and-mortar European retail, at Aldi, Lidl, REWE and Edeka, the sanction mechanisms are less transparent but no less effective. Chargeback penalties of 1 to 5 per cent of invoice value for packaging defects, pallet non-compliance or delivery errors are standard practice across the industry. Flat-rate administrative charges per incident are added on top. For a delivery of 20 pallets with 100 boxes each, a single complaint incident can cost several thousand euros, before accounting for return transport and redelivery. The specific fee structures are governed by individual supplier agreements and are not published.

Pallet Rejection at the Loading Bay

Figure 19 The loading bay as the supply chain's parallel inspection gate. Aerial view of a typical FMCG distribution centre with multiple dock doors operating in parallel: each trailer represents one delivery window, and each window has its own goods-out clerk, its own forklift cycle and its own conformity check against the receiving retailer's specification. The throughput of the entire converter–customer interface is set here, by the slowest of the simultaneously occupied docks. A single non-conforming pallet that ties up one dock for fifteen to thirty minutes does not only delay the affected truck; it shifts the queue at every neighbouring dock and propagates into the dispatch schedule for the rest of the shift. The loading bay is therefore both the last opportunity to catch a packaging defect cheaply and, when it is missed, the first opportunity to lose downstream capacity at scale.

In practice, acceptance rules vary between retailers and between categories. Some chains apply explicit threshold rules drawn from DIN 55440 or internal load-security guidelines (e.g. the EHV quality guidelines in German retail), accepting pallets below a defined damage ratio and rejecting above it; others give the goods-in staff discretion, with blanket rejection the pragmatic default when several outer boxes show visible damage. The rationale where rejection occurs is straightforward: if the outer boxes are compromised, inner damage must be assumed, and individual inspection typically costs more than the pallet is worth to recover. For the supplier, the economic exposure per rejected pallet runs to several hundred to over a thousand euros, even where typically less than one per cent of the products are actually damaged. The practice is less uniform than it is often portrayed, but the cost structure facing the supplier when it occurs is real.

Testing Requirements

Wholesalers and e-commerce companies are increasingly demanding transport packaging tests that comply with standards [42]. In Europe, the relevant standards are ISTA procedures 3K (FMCG supply chains with roll containers and mixed pallets) and 3N (full truck loads in intra-European FTL transport). In addition, DIN 55440 specifies the compressive resistance of packaging in the stack, and EUMOS 40509 tests load stability under horizontal acceleration, with elastic deformation below 10 per cent and permanent deformation below 5 per cent. A packaging that passes the compression test to DIN 55440 can nonetheless fail an EUMOS transport simulation.

4.5 Economic Costs: The Numbers

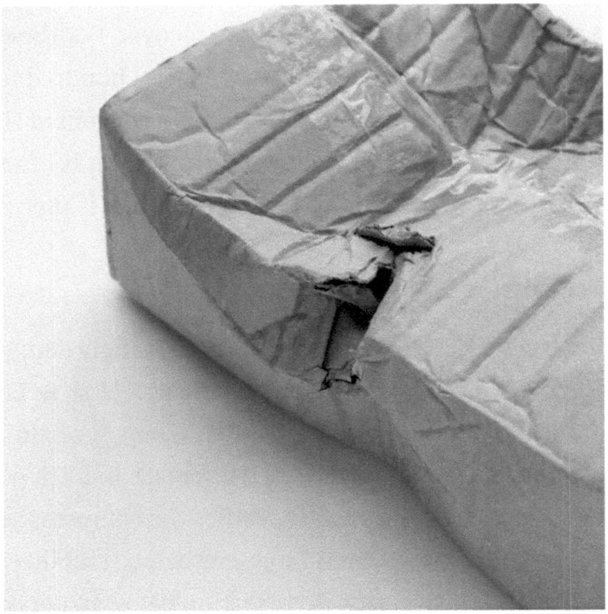

Figure 20 Economic costs in numbers, made visible in a single damaged shipper. The collapsed corner, torn liner and locally crushed flutes shown here are the surface signature of a failure that reaches far beyond the value of the box itself. Industry surveys put the figures into context: the Amcor supplier survey 2020 [36] reports that 73 per cent of FMCG shippers identify packaging damage as a measurable cost driver; Packaging Digest 2024 [57] documents an 11 per cent damage rate at European distribution centres; the EHI Retail Institute 2022 [41] estimates total inventory losses across the European retail chain at approximately 4.1 billion euros per year, of which packaging-related damage is a substantial component. Returnalyze 2023 [47] and DHL eCommerce 2025 [51] add the e-commerce contribution, where return logistics alone double the carbon and cost footprint of every damaged unit. The single box in the photograph is therefore not an isolated anomaly but a unit of currency: each visibly damaged package on a pallet stands for a chargeback at the retailer, a stoppage at the case erector, a return shipment in reverse logistics, or a loss of the end customer — and most often, for several of these at once.

The aggregated costs of packaging failure are striking. Around 10 per cent of all loading units arrive at a European distribution centre with some form of damage [17]. The figure is drawn from DHL's 2025 eCommerce study and refers to inbound units at FMCG and e-commerce distribution centres; rates vary substantially by lane, shipment type and handling regime, and are not directly comparable to heavy industrial B2B, where damage definitions and reporting practice differ.

The volume of e-commerce returns amplifies the effect. In Germany alone, the return rate for clothing stands at 44 per cent [18], and the costs range from 10 to 20 euros per returned item, covering shipping, inspection, and restocking.

The direct losses, however, are only the visible part. The indirect costs far exceed them. Cost Category	Order of Magnitude	Source
Damaged loading units	approx. 10 % of all units	DHL eCommerce 2025 [17]
Returns cost per item (DE)	10 to 20 EUR	Sendcloud 2024 [18]
Proportion packaging-related	34 % of all returns	Returnalyze 2023 [47]
Goods losses in European retail	approx. 4.1 billion EUR per year	EHI 2022 [41]
Lost repeat purchase	73 % do not buy again	Amcor supplier survey 2020 [36]

The final row of the table carries the most serious implications: 73 per cent of consumers do not return to the same supplier after receiving a damaged product [36]. This figure originates from an Amcor supplier survey rather than peer-reviewed research and should be read as an industry indicator, not an academic datum; even allowing for the direction of bias that a packaging-industry publication invites, the order of magnitude is corroborated across multiple consumer studies, and the underlying dynamic – that a damaged delivery is one of the strongest predictors of supplier switching – is well established. This is not a one-off loss but a lasting reputational damage that accumulates over the years.

While the direct cost of a damaged parcel is a few euros, losing a customer can cost hundreds of euros in future revenue. Thirty-four per cent of all returns are packaging-related [47]. The EHI Retail Institute puts annual goods losses in European retail at around 4.1 billion euros [41]. These figures make clear that packaging failure is not a marginal problem but a significant economic factor.

4.6 The Cost Cascade: From One Millimetre to Million-Euro Losses

To summarise the damage path along the supply chain:

Stage 1 – The factory. Fishtailing and dimensional deviation cause stoppages at the case erector. Cost: thousands of euros in hidden OEE losses per shift, misattributed to machine downtime or operator error.

Stage 2 – The pallet. Real stacking strength can lie 50 to 70 per cent below the laboratory value. A weakened MJ makes the box the weakest link in the stack, and a single protruding corner can trigger progressive collapse.

Stage 3 – Transport. Seventeen drop events, vibration, and moisture. Dynamic loading on a glue line designed, at the moment it was laid, for static conditions.

Stage 4 – The loading bay. Pallet rejection. Several hundred to over a thousand euros per rejected pallet for two visibly damaged boxes, regardless of the state of the hundred others behind them.

Stage 5 – The shelf and the consumer. Damaged products, returns, and lost repeat purchases. 73% of customers are permanently lost, and the reputational cost accumulates over time.

Each stage multiplies the cost of the previous one. What begins at the corrugator as 50 µm of washboarding and, at the MJ, leads to a glue line that is 10 per cent weaker, ends as million-euro losses in the supply chain. The dimensional deviation is not the cause; it is the multiplier.

The next chapter broadens the perspective: the environmental costs of packaging failure exceed the economic ones by a significant multiple. A failed box is not only a financial problem; it is a resource problem. The packaging itself typically accounts for only 5 per cent of a product's total climate impact. Protecting the contents is by far the greatest environmental lever. What happens when that protection fails is the subject of the next chapter.

Chapter 5:

Ecological Costs: More Than Just a Damaged Box

Figure 21 Ecological costs, made visible in a single damaged shipper.

Introduction card

A failed box is rarely just a packaging loss. This chapter shows why the ecological cost of failure is usually dominated by the product, the transport, the restart losses, and the recycling contamination that follow. It is essential for readers working on sustainability, PPWR compliance, packaging strategy, or customer-facing quality.

"What is the carbon and waste footprint of corrugated packaging failure beyond the pack itself, and why does PPWR reward precision rather than material reduction in isolation?"

Ecological impact: The ecological cascade has several layers. The first is the embedded CO_2 in the corrugated board itself, about 491 kg CO_2e per tonne. The second is the embedded CO_2 of the product inside the pack, which is usually much larger. In FMCG food packaging, the product-to-packaging carbon ratio can range from 10:1 to 100:1. The third layer is logistics, where returns and reshipments add further transport emissions. Further losses arise from stoppages, start-up waste, cold-chain interruption, and recycling contamination. For damaged food packs, the main ecological loss is often wasted product and wasted transport, not the box alone.

Economic impact: The ecological loss is also a financial loss. Damaged stock must be replaced, returned, written down, or disposed of. These costs include manufacturing value, logistics, administrative effort, lost availability, and customer confidence. In retail supply chains with narrow margins, even small changes in return rate can affect profitability.

Regulatory context: EU Regulation 2025/40, the Packaging and Packaging Waste Regulation, provides the regulatory frame. The chapter explains why a lighter pack that fails more often can be worse than a slightly heavier pack that survives the chain. The relevant aim is the delivered functional unit, not the lowest material input in isolation.

Quality: The chapter connects defect language with sustainability language. It allows quality managers and sustainability officers to describe the same failures using a shared catalogue of consequences.

Limitations: Food-waste attribution varies by product class and region.

The argument is directionally robust, but plant-level decisions need plant-specific CO_2 accounting. Regional averages are useful for orientation, not for audit-ready reporting.

Read in full: sustainability officers, packaging strategists, quality managers, anyone preparing PPWR-compliance documentation.

Card sufficient: executives reviewing the regulatory framework.

Prerequisite: Chapter 4.

Follow-up: Chapter 6, root-cause analysis.

The previous chapter quantified the economic costs: billions of euros in transport damage, chargebacks and lost customers. The economic balance sheet, however, tells only half the story. When packaging fails, value is not the only thing destroyed. The resources already consumed in producing, processing and transporting the contents cannot be recovered, and a further round of resources is drawn in to handle the return, the rejection, and the replacement.

In the medical analogy used throughout this book, we now move from symptoms to prognosis: what happens if the condition is left untreated? The answer is uncomfortable, and it extends far beyond the material value of a corrugated box.

5.1 The Cascade: Box, Contents, Surroundings, Pallet

When a corrugated package fails on the pallet, the damage does not stop at the box. It radiates outward in four stages, each with a wider scope than the one before.

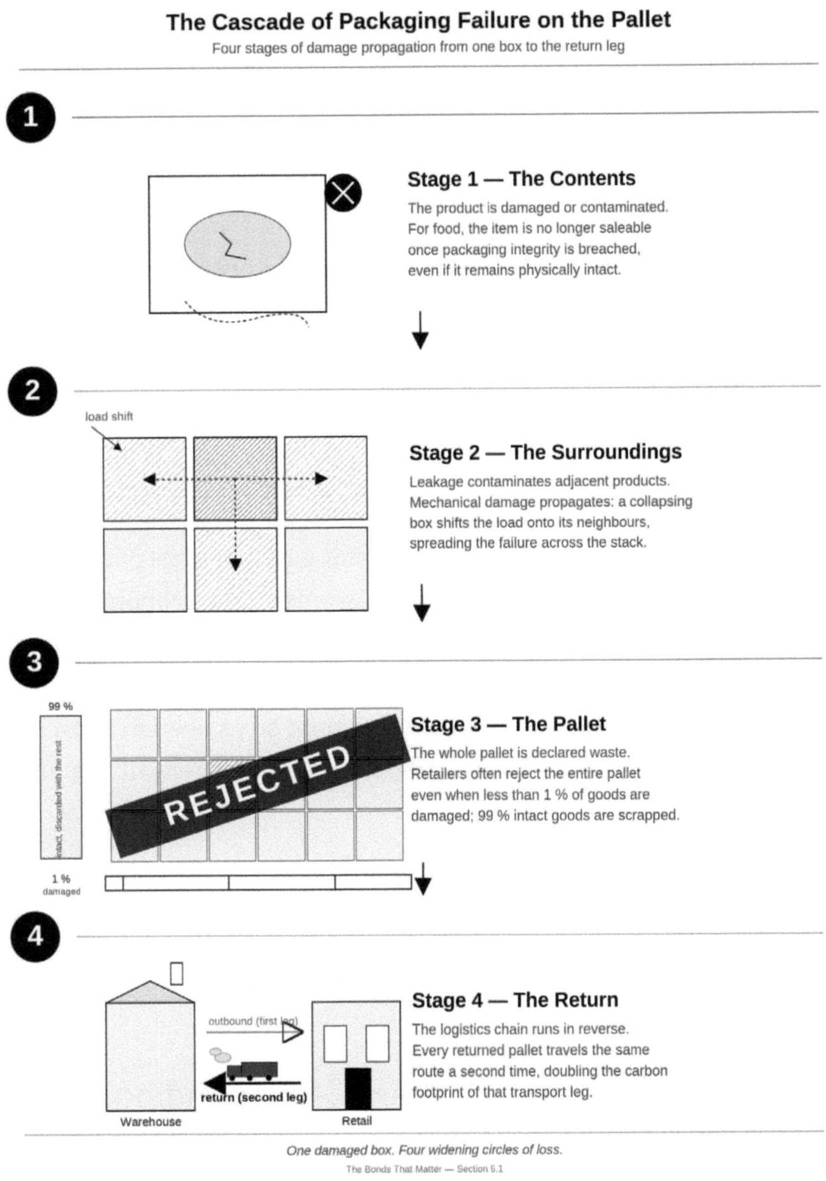

Figure 22 The cascade of packaging failure on the pallet: four stages of damage propagation from a single failed box to the entire return leg, mapped against both the economic and the ecological cost streams developed in

Chapters 4 and 5. Stage 1 (Contents): the product is damaged or contaminated; for food, the unit becomes unsaleable as soon as packaging integrity is breached, regardless of physical product condition (Williams et al. 2012 [61]). Stage 2 (Surroundings): leakage contaminates adjacent products, and a collapsing box redistributes its load onto its neighbours, which were not designed to carry the additional share. Stage 3 (Pallet): retailers routinely declare an entire pallet non-conforming when as little as one per cent of the goods are visibly damaged (DIN 55440, EHV-Richtlinien); the remaining 99 per cent of intact goods are discarded together with the failed unit. Stage 4 (Return): the rejected pallet enters reverse logistics, travelling the same route a second time and carrying the same transport-leg carbon footprint as the original outbound delivery (Singh et al. 2008 [48]; DHL eCommerce 2025 [51]).

Stage 1 — The contents. The first consequence falls on the product itself. If the packaged item is damaged or contaminated, the product is often unsaleable. This is especially true for food if the integrity of its primary packaging is compromised, even if the product remains physically intact. At this stage, the cost is the value of the contents. This is often ten to fifty times the value of the corrugated shipper that failed to protect them.

Stage 2 — The surroundings. Damage spreads from the failed unit to its neighbours. Leakage contaminates adjacent products on the same pallet. Mechanical damage also propagates. When one box collapses, the load shifts to the boxes beside it, increasing their effective load. This exceeds what they were designed to carry. The unit of loss increases as contamination spreads. On a tightly stacked pallet, it often reaches three to five neighbours before stopping.

Stage 3 — The pallet. At the loading bay, the scale of loss grows again. As described in Chapter 4, retailers often reject entire pallets even when less than 1% of goods are actually damaged. The ecological consequences are substantial: 99% of intact goods are discarded or returned, while the 1% that fail. This is the largest single cost step in the cascade — both economic and ecological. It is also the least visible on any conventional damage report, since it is recorded as logistics waste rather than a packaging defect.

Stage 4 — The return. The loss now enters the reverse logistics chain. This involves lorry capacity, warehouse space, handling, and documentation. Every returned pallet follows the same transport route, doubling the delivery's carbon footprint. Return rates in the German clothing market can reach 44%. The added logistics emissions are not a marginal addition to the original shipment. By 2024, European parcel data indicate that reverse logistics costs approach outbound logistics costs (DHL eCommerce 2025 [51]). National greenhouse gas inventories record measurable emissions shares from this process (Thünen-Institut 2023 [59]; Umweltbundesamt 2023 [60]).

The four stages share a common structural feature. The costs scale up by roughly an order of magnitude at each stage. The bare value of the failed shipper is a fraction of a euro. The contents it protects cost ten to fifty times more. Neighbouring units lost in Stage 2 multiply as contamination spreads. Whole-pallet rejection in Stage 3 multiplies the loss by the number of intact units on the same pallet, often by fifty to a hundred. Stage 4 adds reverse-logistics overhead. Damage seen in Stage 1 is a poor predictor of total economic exposure. The cascade, not the single box, is the right unit of analysis.

The ecological structure remains the same, but the units differ. Stage 1 destroys the embodied carbon of the product. For chilled and frozen food, beverages, electronics, and pharmaceuticals, embodied carbon exceeds that of corrugated packaging by one to three orders of magnitude (Section 5.2; FEFCO 2023 [54]; McKinsey 2021 [56]). Stage 2 spreads that destruction across affected neighbours. Stage 3 destroys 99 per cent of intact goods, together with the failed 1 per cent — the highest single ecological cost in the cascade. Stage 4 adds duplicate transport emissions.

The two cost streams, economic and ecological, converge at a single operational lever: the integrity of the shipper as it leaves the loading bay. Every intervention discussed in Part IV, such as stitching for damage tolerance at the joint or waterscoring at the crease, pays off not at Stage 1, where costs are already low, but by preventing the cascade from reaching Stages 3 and 4. At those stages, costs become disproportionate to what the corrugated industry can offset on its own material footprint.

5.2 The Carbon Footprint of Packaging Failure

[Figure placeholder: Carbon-footprint ratios of secondary packaging to packaged contents, across product families]

A common misconception is that packaging is the dominant source of a packaged product's carbon footprint. In fact, packaging typically contributes only five to ten per cent of the total; the remainder arises from agriculture, processing, transport, refrigeration, and preparation.

For corrugated board, FEFCO has quantified the production carbon footprint at 491 kg CO_2e per tonne (2021 data, published 2023, compared with 531 kg CO_2e per tonne in 2018). The emissions break down as follows: paper production (55 per cent), energy consumption in corrugated board manufacturing (20 per cent), upstream transport (10 per cent), and end-of-life incineration (15 per cent) [54].

Figure 23 The five-per-cent trap, made arithmetic. One kilogram of corrugated board saved at the box stage releases roughly 0.5 kg of CO_2 (FEFCO 2021 [54]) — a real and welcome reduction. If that saving causes the e-commerce shipper of a single laptop to fail in transit, however, the device that gets dented, returned, re-shipped or written off carries on its own approximately 300 kg of embodied CO_2 (Dell Latitude product carbon footprint datasheets 2023/24: 312–369 kg CO_2e per device, calculated to ISO 14040/14044). The ecological balance of that single avoided gram of fibre is then negative by a factor of roughly six hundred. The image makes the asymmetry visible: the corrugated box and the laptop appear to be of comparable size, but their carbon footprints differ by three orders of magnitude. Material minimisation that ignores the protective function it minimises does not save CO_2; it relocates it, with interest, to the contents.

When packaging fails, and the contents are discarded, all of the upstream resources have been spent in vain: the agricultural land, the water, the energy used in production and transport, and the refrigeration. The carbon footprint of the lost contents exceeds that of the packaging by a factor of ten to twenty. A decision to make packaging ten per cent lighter at the cost of a two per cent increase in failure rate produces no ecological benefit at all. It produces an ecological loss.

What is decisive here is that, in the vast majority of applications, corrugated board does not serve as the primary packaging of a single product but as secondary or transport packaging. It groups and protects many products that are already in their own primary packaging.

Consider a corrugated tray holding twenty-four beverage cans, a shipping box containing twelve jars of baby food, or an e-commerce carton enclosing a household appliance that is itself already packaged. In each of these cases, the corrugated packaging protects a product value and a carbon footprint that exceed those of the secondary packaging itself by a significant multiple.

Secondary packaging	CO_2 corrugated board	CO_2 contents (total)	Ratio
Tray with 24 beverage cans (retail)	~0.7 kg CO_2e	~7–12 kg CO_2e	1 : 10–17
Carton with 6 × 1 l fresh milk (retail)	~0.5 kg CO_2e	~8.4 kg CO_2e	1 : 17
Crate with 10 kg fruit/vegetables (retail)	~0.4 kg CO_2e	~3–8 kg CO_2e	1 : 8–20
Carton with 12 jars baby food (retail)	~0.6 kg CO_2e	~10–15 kg CO_2e	1 : 17–25
E-commerce shipping box, electronics	~0.3 kg CO_2e	~15–80 kg CO_2e	1 : 50–270
Pharma outer carton, 10 kg (retail)	~0.8 kg CO_2e	~50–200 kg CO_2e	1 : 60–250

The figures in the table are order-of-magnitude indicators, drawn from European product LCAs and FEFCO benchmarks (2021–2023). They are sensitive to three variables above all: (i) the primary product's own CO_2 intensity – electronics and pharmaceuticals are manufacturing-heavy, fresh produce is agricultural, beverages sit in between; (ii) the transport chain's length and temperature regime; (iii) the end-of-life route for the damaged product. Within each product family, the order of magnitude holds robustly across Europe; between families, the range is justified by the table's spread. The argument of this book does not depend on any single row of the table but on the pattern: secondary packaging always offsets its own footprint by a multiple, and therefore any material saving whose price is a higher failure rate produces a net ecological loss, not a gain.

Methodological note. The ratios are computed from cradle-to-grave product LCAs using the GLEC Framework for the transport-related components and EPD-based or FEFCO-benchmarked emission factors for the corrugated packaging. Values from different LCA boundaries (cradle-to-gate, cradle-to-grave, with or without end-of-life credits) can differ by twenty to forty per cent; the bands in the table are chosen to be inclusive of those methodological variants.

The table illustrates the central point: as secondary packaging, corrugated board always offsets its own carbon footprint. Even for relatively low-value products, such as fruit in an open retail crate, the ratio is at least 1:8. For pharmaceuticals and electronics, it ranges from 1 to 250 or more. If the secondary packaging fails, it is not a single product that is affected, but the entire load: twenty-four cans, twelve jars, six bottles. The ecological damage is multiplied by the number of grouped units. In e-commerce, there is an additional dimension: the shipping box must protect the contents not only during palletised transport, but also during individual parcel delivery. Singh et al. (2008; see Chapter 4 [48]) recorded an average of 17 drop events during the journey to the recipient. A single e-commerce carton containing an electronic device has a carbon footprint that exceeds that of corrugated board by a factor of 50 to 270. With a damage rate of three to four per cent in European parcel shipping, this represents a significant, largely avoidable waste of resources.

Protection as a Parallel Environmental Lever

The ecological case for reliable packaging is usually framed in terms of avoided loss. This is accurate but incomplete. Packaging also enables environmental outcomes that no primary packaging, on its own, could deliver at the same efficiency. Dense, regular secondary packaging is the precondition for efficient palletisation and full-truckload transport; where it fails, lanes run with lower load factors and higher emissions per shipped unit. Corrugated outer cases enhance the thermal stability of pharmaceutical and chilled-food shipments, reducing the refrigeration energy required to maintain the cold chain across extended lead times. The secondary packaging that allows a manufacturer to avoid individually over-engineering primary packaging for every unit saves material that would otherwise need to be added at the primary level. These are leverage points in their own right, and they are amplified whenever the secondary packaging performs reliably; they are erased whenever it fails.

5.3 Beyond Carbon: Water, Land, and Forest Origin

A chapter on the ecological balance of corrugated board that confines itself to carbon risks, overlooking the specific footprint of a wood-based material. Water, land use, and the provenance of the virgin fibre have their own weight in the balance, and they follow the same logic as carbon: reliability amplifies the upstream investment; failure wastes it.

Water

Paper and corrugated board production is inherently water-intensive, but the modern European mill is a very different facility from the one the industry is sometimes remembered for. Closed-loop water systems, multi-stage effluent treatment, and aggressive reuse have reduced fresh-water draw at kraft and recycled mills to a range typically cited as 5-15 cubic metres per tonne of finished product, with best-in-class mills below that figure. The dominant share of water passing through the process is recirculated rather than consumed. The argument is not that paper is a low-impact material in the absolute sense, but that its water footprint has been moved, through process investment, into a range where the relative impact of packaging failure, and of the contents it fails to protect, becomes the decisive variable. For food and textile contents in particular, the water embedded in the goods exceeds that embedded in the corrugated packaging by two or three orders of magnitude. The same multiplier logic that governs carbon governs water.

Land Use and the Virgin-Fibre Question

Roughly three-quarters to four-fifths of the fibre in European corrugated board is recycled; the remainder is virgin kraft, drawn from managed softwood and some hardwood forests. The industry's incremental land-use demand is therefore moderate, but it is not zero, and the question of where virgin fibre comes from is not a cosmetic one. Responsibly managed forests are carbon sinks, biodiversity reservoirs, and long-cycle renewable resources; poorly managed or illegally sourced fibre is none of these things. The land-use lever in corrugated is not about using less fibre per box in isolation – that lever is the PPWR's, and its limits are discussed in 5.5 below. It is about ensuring that each tonne of virgin fibre entering the system comes from a source whose management preserves the land's ecological function.

Forest Certification: FSC and PEFC

Forest certification translates the land-use question into a document that the supplier can show, and the customer can verify. Two schemes dominate in Europe: the Forest Stewardship Council (FSC) and the Programme for the Endorsement of Forest Certification (PEFC). Both operate audited chain-of-custody systems that trace fibre from the forest through the mill to the converter and finally to the box on the pallet. Both set criteria for biodiversity, regeneration, indigenous and worker rights, and the exclusion of high-conservation-value forests from harvesting. The two schemes differ in governance and in emphasis – FSC is more prescriptive on social and ecological criteria, PEFC more accommodating of smallholder and family-forest structures – but from the converter's and the brand-owner's perspective, either certificate constitutes a credible assurance that the virgin content meets a defined standard. Adoption in European corrugated is now high: the overwhelming majority of virgin content carries either FSC or PEFC certification. Recycled content participates through the schemes' recycled and mixed credit systems.

Certification is not a substitute for reliability. An FSC-certified box that fails, destroying the contents it was meant to protect, still generates the ecological loss described in 5.1 and 5.2. What certification does is guarantee that the upstream portion of the footprint – the land, the regeneration, the biodiversity – has been paid for in the right currency. The downstream portion, where the package meets the load, the pallet, the truck and the shelf, is the subject of the rest of this book.

5.4 Packaging Failure and Recycling Contamination

Corrugated board achieves recycling rates that few packaging materials can match: above ninety per cent in Europe since 2020 [50], while paper and cardboard as a whole reached 82.5 per cent across the EU [52]. These high rates are a success, but they remain fragile. The principal threat to corrugated recycling is contamination.

The costs of recycling contamination are substantial, even if they are documented less transparently in Europe than in the United States: manual sorting, damage to equipment caused by foreign materials, and the landfilling of rejected loads impose costs running into hundreds of millions of euros annually on European paper mills. Entire recovered-paper batches are downgraded when contaminated bales reduce the fibre quality of the whole delivery.

The Domino Effect in the Recycling Bale

A single carton contaminated with food can render an entire bale of clean fibre unusable. Grease, moisture and organic residues contaminate the surrounding fibres within the bale, and separation is not economically feasible. The result is that the entire bale is downgraded or sent to a landfill.

Wet corrugated board in recycling bales worsens the problem. It can degrade neighbouring dry bales and force mills to downgrade entire shipments. Mills must balance fibre quality with throughput, because contaminated bales reduce prices for Old Corrugated Containers (OCC) for all participants in the recovered-paper market, including converters whose own OCC is otherwise clean.

Packaging failure links directly into this chain. A box that collapses and leaks, whether milk, juice, or canned goods, contaminates not only its neighbours on the pallet but also its neighbours in the recycling bale. The ecological damage propagates through the life cycle, from the loading bay to the mill and back into the next generation of packaging.

5.5 The PPWR Paradox – and Its Resolution

[Figure placeholder: PPWR material-reduction trajectory plotted against process-capability envelope, with the gap where failure risk rises]

The European Packaging and Packaging Waste Regulation, PPWR (EU 2025/40) [26], requires material minimisation: packaging must be reduced to the minimum weight and volume necessary to ensure its functionality. False bottoms, double walls and purely decorative excess are to be prohibited. For e-commerce shipments, empty space in the future may not exceed forty to fifty per cent.

The regulatory intention is right. Less material, for the same functionality, means a lower resource bill. For corrugated board, however, the constraint surfaces an engineering question that the regulation's text does not resolve: reducing material while leaving process conditions unchanged is not a route to lower resource consumption. Lighter liners show more pronounced washboarding (Chapter 2). Narrower tabs increase peel stress disproportionately (Chapter 3). Lower paper weights narrow the tolerance windows for adhesive application, creasing and machine handling. If these process tolerances are not updated to match the lighter material, the failure rate rises, and the ecological savings achieved by a thinner box are more than offset by the ecological costs of the contents it fails to protect. Saving one kilogram of corrugated board saves around 0.5 kg CO_2e. If that saving causes a carton containing ten kilograms of beef to fail, the loss generates 270 kg CO_2e – roughly five hundred and forty times the saving.

The paradox, read carefully, is apparent rather than inherent. It arises only when material reduction is treated as an isolated lever, pulled without adjusting the process that feeds the material. Once that adjustment is made, the paradox dissolves. Material minimisation and process minimisation are two sides of the same regulatory objective: both reduce the resource bill; both depend on the protective function remaining intact; both require converters to treat the line, the corrugator, the folder-gluer and the packing station as a single system rather than as independent cost centres.

Read this way, the PPWR is not a constraint to be resisted but a prompt for the kind of smart process management that the final part of this book sets out in detail. Stitching (Chapter 10) and Waterscoring (Chapter 11) are two interventions that deliver the same structural integrity with lower material content; the integrated concept is developed across Chapters 12 and 13. The regulation defines the direction of travel. Engineering defines how far one can travel in that direction without losing the protective function that justifies the packaging in the first place.

5.6 Interconnected Production Systems: The Ecological Multiplier

Chapter 4 quantified the economic costs of line stoppages. The ecological costs follow the same logic and are no less serious.

Product spoilage. When a food or pharmaceutical line comes to a halt because cartons jam, temperature-sensitive products may remain on the conveyor while the cold chain begins to degrade; in pharmaceuticals, entire batches may become unusable. The direct cost of a single incident in the food industry ranges from 2,000 to 80,000 euros, depending on batch size, the duration of the stoppage, and the product's temperature sensitivity.

Start-up losses. After a stoppage, the first products off the restarted line often fail quality requirements and are discarded. This waste is a function of the restart itself and is incurred independently of the cause of the stoppage.

Energy drawn without output. The line infrastructure – cooling, ventilation, compressors, lighting – continues to draw power during a stoppage without producing anything. The idling fraction of the line's electricity demand is not trivial: in the food and beverage industry, motor-driven systems account for more than seventy per cent of electricity use, and a significant share of that load continues to run whether the line is producing or stopped.

The ecological multiplier operates in all directions simultaneously: material is wasted, energy is consumed, emissions are generated, and no saleable product leaves the line. The trigger is a dimensional issue in a corrugated outer case.

5.7 The Greatest Ecological Contribution of Packaging Is Its Protective Function

Figure 24 Protective function as the primary environmental lever. The "Fragile" marking and the careful, gloved handling pictured here represent the visible end of a chain of decisions whose environmental weight far exceeds the carbon footprint of the corrugated box itself. As Section 5.2 has shown, the embodied CO_2 of a typical FMCG product, electronic device or pharmaceutical exceeds that of its corrugated shipper by between one and three orders of magnitude. A box that arrives intact protects an asset whose ecological cost has already been incurred; a box that fails in transit converts the entire upstream investment, raw materials, manufacturing energy, refrigeration, outbound logistics , into waste, irrespective of how recyclable the corrugated fibre on the floor of the recycling plant happens to be. The protective function is therefore not a secondary attribute alongside material efficiency; it is the parallel environmental lever, equal in weight to material minimisation and in many product categories considerably greater. The handling marking on the carton expresses an intent. The packaging structure has to deliver on it, and the FRAGILE label on its own , as Chapter 4 noted for

the last-mile leg , functions more reliably as an after-the-fact disclaimer than as a sorting instruction the supply chain actually obeys.

The packaging itself accounts for only five to ten per cent of a product's total climate impact. The greatest ecological contribution of packaging, therefore, lies not in its own material efficiency but in its protective function. A corrugated box that reliably protects its contents is ecologically more valuable than a lighter box that fails more often, by a margin that the arithmetic of the product-LCA table makes inescapable.

Recycling contamination caused by packaging failure puts at risk a success rate of more than 90%. Certification schemes – FSC and PEFC – guarantee the upstream provenance of the virgin fibre, but they neither contribute to nor deduct from the reliability of the finished box on the pallet. The PPWR requires material reduction, and correctly so; the regulation also recognises that the protective function must be maintained. The reconciliation lies not in contradiction but in innovation: processes that deliver more protection with less material.

That is precisely the subject of Parts III and IV of this book: first, the systematic analysis of causes (Chapter 6), then the two interventions, Stitching (Chapter 10) and Waterscoring (Chapter 11), that point the way towards packaging that is both more sustainable and more effective. The conclusion is the same in ecological and in economic terms: less material, less adhesive, better boxes.

Part III – Root Cause Analysis (Diagnosis)

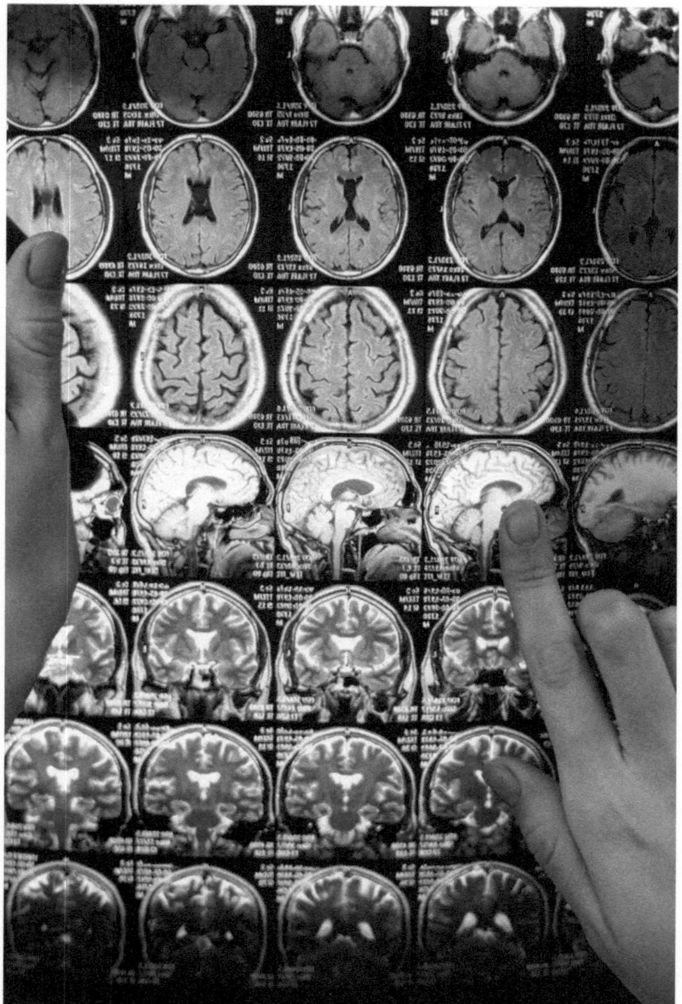

Diagnosis bridges visible symptoms and root causes. Part III traces each symptom identified in Part II back to its underlying mechanism. It examines the overall causal chain and the two families of signatures recognised on the finished blank: folding variation and the secondary fracture that sometimes forms beside the crease. The section concludes with the adhesive application defects that even the most precise glue application cannot wholly prevent.

Chapter 6:

Systematic Root Cause Analysis – From Symptom to Source

Introduction card

Visible defects do not usually have single causes. This chapter introduces a diagnostic method that separates symptom, mechanism, and upstream source. It is the bridge between the failure cascade in Part II and the interventions in Part IV. It is especially relevant for audit managers, quality teams, and continuous-improvement programmes.

"*How do you trace a manufacturer's-joint defect back to its origin across a multidimensional process chain, when most lines run most of the time, and the same defect can have several upstream causes?*"

Quality: The chapter uses a three-step diagnostic framework: symptom, mechanism, and upstream cause. A symptom is what appears on the blank or in the pack. A mechanism explains how the defect was produced. The upstream cause identifies where correction must take place. This distinction prevents teams from correcting the visible symptom at the wrong station.

Process: Variation is treated as the normal state of production. The task is not to eliminate all variation, but to understand which parts can be controlled and which must be compensated for. Machine settings such as pressure, speed, and dosage can be adjusted. Substrate variability, climate, and residual upstream defects cannot always be controlled in real time.

Operational efficiency: Most production defects are not one-cause events. They usually combine one primary mechanism with several secondary contributors. A simple one-cause-one-effect audit often misses this interaction and directs corrective action to the wrong point. The multidimensional matrix introduced here is the basis for a more effective audit.

Process integration: The matrix prepares the reader for later interventions. Stitching does not remove variation, but makes the joint less vulnerable to it. Waterscoring narrows the variation at the crease where it matters most. Together, they show why the book moves from parameter optimisation to system robustness.

Limitations: The framework names variables, but it does not measure them. Quantification follows in later chapters. Plant-specific implementation requires plant-specific data.

Read in full: process engineers, quality managers, audit managers, anyone running a continuous-improvement programme on a folder-gluer.

Card sufficient: executives reviewing the diagnostic framework.

Prerequisite: Chapter 5.

Follow-up: Chapter 7, causes of folding variation.

The previous chapters described each mechanism separately. They covered washboarding and warp at the corrugator (Chapter 2), their compounding at the manufacturer's joint (Chapter 3), the cascade from converting line to retail shelf (Chapter 4), and the underlying ecological costs (Chapter 5). What remains missing is the connecting tissue. Symptoms alone do not yet provide a diagnosis. A diagnosis requires showing how mechanisms interact, why optimising individual parameters is not enough, and, above all, where to place an effective intervention.

This chapter provides that diagnosis. It brings findings together into a clear picture. It shows why the conventional approach—optimising parameters at the corrugator and at the folder-gluer—has reached its limit. This answer prepares the way for Part IV, which focuses on system-level robustness instead of tighter machine-level settings.

6.1 The Causal Chain: An Overview

Figure 25 The causal chain from corrugator to shelf, in ten links. Each link conditions the next, but the chain is not a linear sequence that could be repaired link by link. Link 1, downgauging, reduces the fibre mass that everything downstream depends on. Links 2 and 3 amplify the resulting deformation into geometric asymmetry at the bonding interface. Links 4 to 6 trace the adhesive into a contact zone whose conditions are already set by the flute geometry and washboarding before the glue arrives. Link 7 captures the joint at its most exposed moment — five to fifteen per cent of final film strength against the corrective force of the side joggers. Link 8 adds the spring-back of warp and crease stiffness as peel and cleavage on the still-fresh bond. Link 9 is the manufacturer's joint failure itself; Link 10 the cascade of consequences along the supply chain. The footer states the structural reading: this is a system with feedback loops, not a line that can be cut once. Moisture acts on washboarding and warp at the same time. Downgauging sharpens the washboarding and narrows the adhesive tolerance window. Tab width determines how much redundancy the system retains, and the PPWR removes precisely that redundancy in the name of material efficiency. The interaction matrix that follows in Section 6.2 makes the same argument quantitative; the closing line of the figure — "Strengthen one link and the others breathe; weaken one, and the whole chain tightens around it" — anticipates the system-level conclusion of Section 6.4.

The causal chain from the corrugator to the retail shelf has ten links. Each condition is the next. Alone, none are enough to cause failure. But together, they show how one design choice—using lighter paper—leads to a rejected pallet.

Link 1 – Downgauging. Using lighter paper grades and thinner liners reduces the fibre mass per unit area (grammage). While accounting for the reduced material is straightforward, understanding the mechanical effects—such as decreased strength and increased flexibility—is more complex.

Link 2 – Localised deformation. Reduced fibre mass increases local deformation at flute-tip bonds, intensifying washboarding. The first sign is the liner's loss of surface flatness.

Link 3 – Asymmetric contact. Corrugated geometry causes uneven contact, worsened by washboarding. Lightweight liners deflect more, increasing adhesive interface gaps.

Link 4 – Adhesive application. At the FFG, adhesive meets a modulated surface. Contact systems wipe adhesive off peaks; non-contact systems contend with differing porosity.

Link 5 – Strike-in asymmetry. Flute-tip densification slows strike-in, keeping adhesive fluid and easily displaced during joining.

Link 6 – Joining. During joining, two surfaces are pressed together with contact conditions already dictated by the flute geometry and by washboarding. The adhesive is expelled from the contact zones to where it is least needed.

Link 7 – Green strength and corrective forces. The PVAc (polyvinyl acetate) dispersion adhesive builds strength slowly. At the delivery stacker, it has reached only 5–15 per cent of its film-forming (fully set) strength, yet it already carries green load (the initial load applied before full setting). At the same moment, side joggers (mechanical guides) exert corrective forces on the bundle; when set too tight, they impose sustained shear (sliding force) on the still-fresh joint (see Section 3.1, Phase 5).

Link 8 – Spring-back. Warp (curvature distortion) and the bending stiffness at the crease line generate an opening force on the fresh glue line. The load is applied as peel (separation along the surface) or cleavage (splitting), which are the two most unfavourable modes for a film-forming adhesive.

Link 9 – Manufacturer's joint failure. The MJ (manufacturer's joint) can fail partially or completely. The box then shows dimensional deviations (size errors), fishtailing (misalignment), or springs open at the delivery stacker. The production line has produced its first defective blank (an unusable box); by the time it is recognised, several thousand more are already on the pallet.

Link 10 – Cascading loss. One defective box leads to stoppages, pallet failures, chargebacks, recycling issues, and loss of end customers.

This chain is not a simple sequence where fixing one defect would solve the problem. It is a system with feedback loops and mutual reinforcements. Moisture influences washboarding and warp simultaneously. Downgauging increases washboarding and narrows adhesive application windows. Tab width sets redundancy, and the PPWR [26] removes it in the name of material efficiency. The next two sections will show why such a system cannot be fixed by adjusting parameters one by one.

6.2 The Interaction Matrix

The matrix below shows how eight main factors interact. Each cell shows the effect the row factor has on the column factor. The scale is coarse—five steps, not a number—because the evidence is qualitative. These classifications are based on a synthesis of primary studies across Chapters 2–5. Controlled factorial experiments are not possible in systems this complex.

Reading the matrix: ++ strong amplification · + moderate amplification · o neutral · - moderate relief · -- strong relief. Diagonal cells carry - to indicate that the factor's self-interaction is not informative and is omitted.

	WB	Warp	Liner	Tab	Crease	Moist.	Gram.	Pallet
Washboarding	–	+	+	++	+	++	++	+
Warp	+	–	+	++	+	++	+	++
Liner type	+	+	–	+	0	++	+	0
Tab width	++	++	+	–	+	+	+	+
Crease quality	+	+	0	+	–	+	+	0
Moisture	++	++	++	+	+	–	+	++
Grammage ↓	++	+	+	+	+	+	–	+
Pallet cond.	+	++	0	+	0	++	+	–

The matrix highlights one telling feature. Most off-diagonal cells amplify rather than relieve. Almost no pairs are neutral, and none offer relief. So, if one factor worsens, almost every other factor gets worse too. The system is unstable and tends to amplify problems rather than correct them.

Moisture and grammage columns are especially informative. Both show mostly amplifying effects. Moisture strengthens washboarding and warp, reduces BCT, and speeds glue line damage. Lower grammage increases washboarding, narrows adhesive tolerance, and raises moisture sensitivity. Moisture and downgauging together are, by the matrix, the worst pairing.

6.3 Variation: The True Enemy

In daily work, optimisation focuses on average values. Operators target the ideal adhesive amount, crease depth, and viscosity. But the issue is not the mean—it is the variation around it.

A systematic deviation—like always using too much starch at the corrugator—can be fixed with one adjustment. Variation is different. Here, the value changes from blank to blank. No single setting can address this. Operators set up for the average. Every blank then varies in unpredictable ways.

Flute phase position as a case study

[Figure placeholder: Phase relationship between creasing tool and flute valley varies across the web and over time]

Nagasawa and Komiyama (2012) [29] studied how the crease tool's position affects the spring-back moment. Their formula $\Delta M\_peak = 2.21 \times (e/\lambda) + 0.34$ links how far the crease is from the flute valley (eccentricity e) to the flute pitch (λ). The further from the valley the crease sits, the higher the spring-back at the glue line.

[Figure placeholder: Inline creasing in the machine direction produces an unavoidable scatter of crease qualities across the web]

Nica et al. (2020) [30] confirmed this effect for inline creasing. Since the crease is applied in the machine direction, its phase with the flute geometry changes across the web and over time. This causes unavoidable scatter in the crease quality. Tip, shoulder, and valley alternate unpredictably along the web.

Even if every corrugator and FFG parameter is set just right, flute phase position still causes random variation in spring-back. This scatter cannot be eliminated by conventional methods. Each blank ends up with a different crease quality and a unique risk of failure.

Flute crushing: residual deformation as a hidden variable.

[Figure placeholder: Elastic recovery of a crushed flute leaves flattened tips; the bond geometry after recovery is not the original geometry]

Garbowski et al. (2021) [4] studied board crushing during converting. The folder gluer's belt pressure compresses the flute tips. Some deformation recovers elastically when pressure is released. Some is permanent. Flute height always decreases after conversion, reducing stiffness and stacking strength.

Crushing causes variation, not just less stiffness on average. Residual deformation is not uniform. It depends on local belt pressure, fluting moisture, flute type, and liner stiffness. Along one blank, crushing varies flute to flute. Between blanks, it varies with belt position and belt wear.

Even where flute height partially recovers, the recovered geometry is not the original geometry. The flute tips are flattened, the contact area at the bonding point has shifted, and the local stiffness profile of the board is altered in ways that cannot be predicted solely from the machine settings.

Jamsari et al. (2020) [8] confirmed this picture through bending and edge-crush tests at different pre-crushing levels. As long as the crushing stays moderate, the drop in bending stiffness tracks the reduction in calliper in a roughly linear fashion. Beyond about fifty per cent flute-height loss, however, the experimental edge-crush force collapses far more steeply than any of the analytical or finite-element models predict. The board enters a regime in which the residual geometry no longer carries load as the intact profile did, and the safety margin the converter assumed at the design stage has quietly evaporated.

The consequences for the MJ are twofold. A crushed flute tip produces a different contact condition during bonding than an intact flute: the adhesive wets a broader, shallower zone, and the bond geometry changes accordingly. The reduced flute height also lowers local board stiffness and redistributes the stress under pallet load. Both effects vary from blank to blank, and neither is visible from the outside.

Principal variation drivers and their controllability

Variation driver	Cause	Controllability
Flute phase position	Crease position vs. flute geometry	Not controllable
Washboarding amplitude	Starch add-on at corrugator, liner moisture	Partially controllable
Flute crushing (residual)	Belt pressure, fluting moisture, flute type, liner stiffness; partial elastic recovery leaves flattened tips	Barely controllable
Warp direction and magnitude	Differential stresses, storage climate	Partially controllable
Liner porosity	Batch, liner type, moisture history	Barely controllable
Adhesive viscosity	Temperature, shear, pot life	Controllable
Pallet position	Deck board gaps, overhang, stacking pattern	Barely controllable

The table makes the fundamental problem visible at a single glance. Most variation drivers are not controllable, or only partially so. The flute phase position is physically unavoidable. Liner porosity varies from batch to batch.

The position of a given box on the pallet is beyond the manufacturer's reach. Conventional parameter optimisation can act only on the controllable factors, and those are in the minority. Chapter 7 returns to this catalogue in more detail (Section 7.7 develops the mechanism behind each row); for the argument of this chapter, the distribution of entries in the right-hand column is enough.

6.4 Why Conventional Optimisation Reaches Its Limits

The conventional response to a failing MJ is familiar: optimise the adhesive quantity, adjust the viscosity, increase the belt pressure, and re-check the creasing. Setting optimisation works, up to a point. Three systemic limits prevent it from ever resolving the problem permanently.

Limit 1 – Variation. As Section 6.3 showed, the most important variation drivers are not controllable. Setting optimisation acts on mean values; it cannot narrow the scatter. The optimum operating point is a compromise that holds for the average value and fails at the tails of the distribution, which is precisely where the rejected blanks live.

Limit 2 – Interactions. The matrix in Section 6.2 shows that improvements in one area are routinely offset by deterioration elsewhere. More adhesive compensates for washboarding, but increases open time and the risk of strike-through on porous liners. The parameter space is multidimensional, and the optimum lies at a different point for every linear type and every climate.

Limit 3 – External drivers. The PPWR [26] demands material optimisation: narrower tabs and lighter grammages. Warehouse humidity is not controllable. These external drivers narrow the parameter space within which conventional optimisation operates. The room for manoeuvre is steadily shrinking, independently of how skilfully the operator works inside it.

The lesson is straightforward. The corrugated industry does not need better settings; it needs approaches that reduce variation and raise system robustness. Those are qualitatively different objectives. A robust system tolerates the scatter that a variation-reducing system would still try to suppress; a variation-reducing system tolerates the scatter that an optimum-seeking system would try to move. The conventional toolkit was built for the third of these, and the third no longer suffices.

6.5 A Preview: How the Matrix Changes with Stitching and Waterscoring

The matrix in Section 6.2 describes the field as it presents itself today: almost every pairing amplifying, not one pairing relieving. The question that naturally follows is how the same matrix would look if the two interventions developed in Part IV were introduced into it. A full answer is the business of Chapter 13, where the integrated concept is drawn together; a short preview, however, belongs here, because it clarifies why the two interventions are the right candidates.

Waterscoring changes the crease-quality row and column. Local conditioning reduces bending stiffness at the crease line by a factor of three to four, which in turn reduces the spring-back force acting on the fresh glue line. Because the conditioning is local and dosed, the effect is largely independent of the flute phase position: whether the crease lands on a tip, a shoulder, or a valley, the conditioned zone softens the transition. Several amplifying cells in the crease row shift toward neutral, and the coupling between crease quality and moisture, currently one of the stronger amplifiers, is substantially dampened.

Stitching alters the joint's response. A segmented glue line interrupted by mechanical fasteners is damage-tolerant rather than strength-maximising: a localised defect in one segment does not propagate along the seam. Several of the amplifying interactions in the washboarding, liner-type, and grammage rows therefore no longer translate into MJ failure, even when the cells themselves remain amplifying at the material level. The matrix of material interactions is unchanged; the matrix of failure consequences is transformed.

The two interventions do not remove variation – nothing in the field can – but they decouple variation from failure. That is the structural shift Part IV will develop in detail. Stitching delivers damage tolerance at the joint; waterscoring relaxes the boundary condition at the crease. Together, they convert an amplifying field into one that tolerates, and in places relieves, the disturbances that conventional optimisation could only react to.

Part III has diagnosed the system; Part IV now treats it. Chapter 10 develops the concept of stitching as a principle of damage tolerance. Chapter 11 develops waterscoring as a local conditioning of the crease. Chapter 12 takes waterscoring beyond the folder-gluer to the packing line. Chapter 13 reassembles the diagnosis and the treatment into a single integrated concept.

Chapter 7: Causes of Folding Variation – Why the Spread Decides More Than the Mean

Introduction card

Average crease quality does not determine field performance. The weakest crease often does. This chapter explains why folding variation arises, how it appears in the bundle, and why random scatter must be treated differently from systematic deviation. It is especially useful for process engineers and quality leaders building practical control systems.

"Why does the spread, not the mean, of crease quality decide line performance, and why is the weakest crease, not the average, what determines pack performance under stack loading?"

Quality: Fold quality is influenced by several sources of variation. Flute phase changes along the crease line. Washboarding changes with starch add-on and moisture. Liner porosity varies by batch. Moisture varies between pallets, blanks, and times of day. Crease pressure changes with rule wear and tool-mounting tolerances. Together, these factors make each point on a crease part of a distribution, not a single stable value.

Operational efficiency: A box at the bottom of a pallet column does not fail according to average crease strength. It fails where the weakest crease or panel section can no longer hold the load. This explains why field behaviour can differ from laboratory averages and why interventions that reduce scatter can be more valuable than interventions that only raise the mean.

Diagnosis: The chapter teaches the reader to distinguish systematic deviation from random scatter. A systematic deviation may justify correction at the folding bar, creasing pressure, or creasing wheel. Random scatter points upstream to pallet moisture, warehouse climate, cut register, flute phase, infeed skew, or crushing. Confusing the two is one of the most common shop-floor errors.

Process: The chapter provides the basis for SPC at the folder-gluer. If the field metric is weakest-crease performance, control systems must monitor the lower tail of the distribution, not only the mean.

Limitations: This chapter analyses folding variation. It does not yet propose the interventions. Those follow in Part IV.

Read in full: process engineers, quality managers preparing audits, shift leaders, anyone designing crease-quality SPC.

Card sufficient: executives reviewing the variance-control argument.

Prerequisite: Chapter 6.

Follow-up: Chapter 8, the secondary fracture.

Chapter 1 defined the two surface signatures of folding failure: the off-dimension box and the fishtail. Chapter 3 followed them to the focal point where they converged. Chapter 6 placed them inside a system that amplifies rather than self-corrects. The present chapter reads them back along the line to the mechanisms that produced them.

In this chapter we follow the fold signatures back through four causal families: the machine (lateral offset, skew and cut-phase drift at the creasing station), the pallet (non-uniform moisture across the incoming stack), the upstream cut (the drift of the rotary shear and its written record on the corner slit), and in-press crushing (the nips and belts that pre-damage the blank before it ever reaches the creasing rule). Each family leaves a different signature on the finished bundle; each calls for a different response. Confusing one family for another is the most common diagnostic failure on the shop floor, and the one that does most damage to line productivity.

The craft of the converter does not lie in eliminating variation; variation is the default state of every line. The craft lies in reading variation, narrowing the band where it matters, and compensating for what remains. The sections that follow arrange the four families in the order in which their signatures become legible: first at the creasing station itself, then on the pallet feeding it, then on the corner slit written by the corrugator, and finally in the crushing pattern left by the machines along the way.

7.1 The Creasing Station and the Crease Line

Lateral offset at the infeed

The first source of variation is the lateral offset. If the blank enters the creasing station even a few millimetres off-centre, the whole crease line shifts out of phase with the flute. An offset invisible to the eye at the machine can significantly alter the pressure distribution along the crease. At one position, the creasing wheel presses into a flute trough; at another, it sits on a flute tip. The fold closes unevenly under the packer: a split shows on one side, while the other springs back.

The offset is often blamed on the side guides, but the root cause can sit further upstream. A worn feed belt or a slipping vacuum table produces the same effect, as does a blank with a slightly curved leading edge from the upstream die cutter. Each feeds the same tool with a different initial position: the board looks identical, yet the behaviour does not.

Skewed feed

A few millimetres of rotation are enough to let the crease line wander continuously across the flute wave. At the leading edge, the crease may meet a flute tip; at the trailing edge, it may sit in the trough. The crease quality varies along the fold, with consequences extending to the joint.

Skew is hard to read on a single sheet. It shows up best across a bundle when the top edges of the top ten flats are laid against a straight edge. A jogger, a laser line or a simple steel rule along the bundle side all reveal it. These are things the machine-guiding devices missed.

Inline measurement of lateral offset and skew (2024 generation)

For a long time, offset and skew at the folder-gluer infeed were detectable only by their consequences: a drifting crease line, a widening gap, a fishtail on the pallet. The 2024 generation of optical register monitoring has begun to close that gap. Modern systems project a line-scan camera across the full working width and track the leading and trailing edge of every blank in real time, resolving lateral offsets to better than 0.3 mm and skew angles below 0.05°. Garbowski (2024) [5] surveys the integration of such sensors into the FFG control loop; Rogalka et al. (2024) [14] report a convolutional network that classifies deformed blanks in situ from a single overhead frame, with accuracy sufficient to trigger operator alerts ahead of the creasing wheel.

The transition from "see the fishtail on the pallet" to "flag the skew at the infeed" is the single most consequential change in shop-floor diagnostics since the introduction of washboarding measurement. It does not replace the bundle reading developed in Section 7.4; it complements it by closing the latency between cause and evidence from hours to seconds.

Cut position relative to the flute

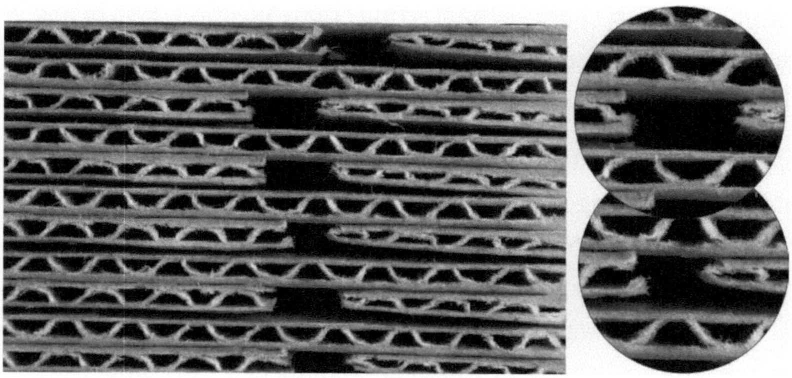

Figure 26 Cut position relative to the flute, written into the bundle. Side view of a stacked bundle of knock-down flats with the corner slits aligned for inspection (main image), and two close-up details (right) of the same edge at higher magnification. Each flat in the bundle terminates at the slit on a different point of the flute wave: some end on a complete flute peak, some halfway through the flank, some in the open trough between two flutes. The pattern is not random handling damage; it is the direct fingerprint of the cut register at the corrugator, repeated one flat at a time and laid down in the order in which the flats left the rotary shear. Read along the bundle, the slit pattern reveals one of three diagnostic signatures (Section 7.4): a repeating pattern, in which every slit shows a similar flute position, indicates a cut register that drifts slowly and within narrow bounds — the upstream machine is holding its phase reference and the creasing station has a predictable starting point. A wandering pattern, in which the flute position drifts smoothly from complete to cut to complete across the bundle, indicates register drift in the upstream cutting and routes the corrective action back to the corrugator team. An erratic pattern, in which no two slits look alike and the flute position jumps from one flat to the next, indicates a mechanical disturbance: a worn knife, a slipping feed, a vibration in the rotary shear. The cause sits not at the folder-gluer but at the cut, and the bundle records it with millimetre accuracy. The same upstream variation that the bundle makes legible at the loading dock is what drives the flute-phase scatter quantified by Nagasawa and Komiyama (2012) [29] and Nica et al. (2020) [30] — and,

downstream, the parallel-offset and fishtail signatures revisited in Sections 7.5 and 7.6.

The third source happens upstream at the corrugator or the rotary shear. The blank is cut from the board at a set position relative to the flute. Cutting a few millimetres off changes the starting crease position. Two blanks from the same pallet may thus carry their creases on entirely different flute positions. The stack looks identical; the creases do not.

The operator at the folder-gluer sees identical blanks in the feeder. The operator at the packing line sees two different boxes. The root sits at the upstream cutting register, not at the creasing tool. A perfectly stable cut position is not something the line can deliver: the cutting register drifts with speed changes, knife wear, splice joints in the liner and temperature. The drift is normal. Whether the downstream tooling can still compensate depends on its rate and amplitude.

The two signatures at the fold

The three sources produce two clear signatures at the joint, and the difference matters because the fixes differ. The first signature is the parallel offset. The crease line runs straight, but sits at the wrong flute phase over its whole length. Every point of the crease faces the same flute position – tip, flank or trough – and the deviation is consistent along the fold.

Nagasawa and Komiyama (2012) [29] and Nica et al. (2020) [30] describe how a crease that meets a flute tip deflects differently from one that meets a trough, with a fold-point shift that can reach a substantial fraction of a flute pitch. On the finished box, a parallel offset appears as a fold that sits a few millimetres too far in or too far out, with the two edges of the manufacturer's joint running parallel to each other. The gap is uniformly too wide or too narrow, not twisted. The root cause is the cut position at the corrugator or a persistent lateral offset at the folder-gluer.

The second signature is the skewed crease – the mechanism behind fishtailing. Here, the crease line meets the flute at an angle rather than perpendicularly. Because the resistance of the flute structure is direction-dependent, the creasing wheel follows the path of least resistance. The crease wanders from tip to flank to trough along the fold, and the line itself begins to meander. When the side panels are erected, the fold angle varies along the crease. One end of the joint opens wider than the other, and the edges splay outward like a fish's tail. Wade et al. (2023) [18] and Jamsari (2020) [8] both document this direction-dependent response of the corrugated structure.

The remedies track accordingly. A parallel offset is corrected by chasing the cut register at the corrugator or by tightening the lateral alignment at the folder-gluer. A skewed crease is corrected by reducing the angle between the creasing and flute axes, which means addressing skew at the infeed rather than at the creasing pressure. A shift leader who applies deeper creasing pressure to a fishtail worsens the crushing without correcting the skew.

7.2 Moisture Across the Pallet Before Converting

Where the moisture damage really happens

Moisture is often discussed in the context of the finished bundle, but by then the blank has already been creased, folded and glued. Whatever the moisture does to the flat at that stage affects storage stability and the box's behaviour at the customer, not the quality of the crease itself. The crease was formed hours or days earlier, at the folder-gluer, from a pallet of sheets or blanks that carried their own moisture signature into the machine. That is where moisture variation rewrites the crease geometry and the fold behaviour.

The pallets that matter for this discussion are the pallet of corrugated sheets leaving the corrugator and the pallet of die-cut blanks arriving at the folder-gluer. Both sit in the warehouse or in the work-in-progress buffer. Both exchange moisture with the ambient air on their outer layers. The core sheets remain sealed. When the sheets enter the converting line, they bring a distribution of moisture contents with them, and the converting machines react accordingly.

The pallet is not a uniform body.

[Figure placeholder: Moisture gradient across a pallet: outer sheets exchange with the ambient, inner sheets hold their state]

A pallet of corrugated sheets looks like a uniform block. A closer measurement tells a different story. The moisture inside varies from the centre to the edge, from the top sheets to the bottom, from the wrapped side to the open side. The difference can reach several percentage points between a sheet at the core and a sheet three layers in from the exposed edge. Bronlund et al. (2013) [2] and Modzelewska (2006) [10] have both documented steady-state moisture transport and the hygrostability of corrugated board under typical warehouse conditions.

The cause is simple. The outer sheets exchange moisture with the ambient air; the inner sheets are cushioned against the warehouse climate. Overnight, the outer layers take up or release water. The inner layers hold their state. When the pallet is opened at the folder-gluer, and the sheets are fed in one after the other, two populations of board pass through the creasing wheels: the drier core sheets and the wetter or drier edge sheets. Each population meets the same tool with different mechanical properties.

Why do the edge sheets behave differently?

The edge sheets were first exposed to the warehouse climate. A humid warehouse pushes moisture into the outer sheets. Flutes swell slightly. Liner fibres lose stiffness. The crease that was correct for the dry core sheets now sits on a softer substrate, and the creasing pressure calibrated to the dry condition is too high for the damp edge. Crushing deepens, spring-back falls, the crease line widens. A dry warehouse removes moisture from the outer sheets. Fibres become brittle, the crease meets a substrate that cracks rather than folds, and secondary fracture lines develop next to the primary crease. Cornaggia et al. (2023) [3] and Szewczyk and Głowacki (2014) [15] document the temperature- and humidity-dependent mechanical properties in detail.

The pallet geometry plays its part. A corner sheet exchanges moisture on two sides, a side sheet on one, a core sheet on none. The exchange rate scales with the exposed area. A tall pallet with a narrow footprint has more edge than a short pallet with a wide footprint. A stretch wrap of polyethene slows but does not stop the exchange. A perforated wrap offers no barrier at all. A pallet that has sat overnight in a heated production hall and is moved into a cold warehouse starts a transient exchange on every face.

The practical signature at the folder-gluer

The operator at the folder-gluer who sees two different fold behaviours on consecutive sheets is not looking at an erratic machine. They are looking at two populations of sheets that sat in different positions on the same pallet. The blame falls easily on the creasing tool or the adhesive. The real cause sat in the warehouse aisle for twelve hours, exchanging water with the night air.

The effective remedies are: equalise the warehouse climate, minimise the time between corrugator and folder-gluer, keep pallet wraps intact during sensitive runs, and, above all, maintain awareness that the pallet is never as uniform as it appears.

7.3 The Two Kinds of Crushing

The story so far risks leaving the impression that every form of crushing in corrugated board is damage. The literature tells a more precise story. At a specific location and with a specific tool, crushing and ply delamination are not defects; they are the mechanism that enables folding. The distinction is essential: the desired crushing at the crease line is the cure, the undesired crushing elsewhere is the disease. The folder-gluer process depends on distinguishing the two.

Why must the crease be crushed?

A flat corrugated panel is stiff because the two liners and the fluting medium act together as a composite beam. The outer liners carry tension and compression at the extreme fibres, the fluting carries shear across the thickness, and the adhesive bonds at the flute tips keep the plies coupled. The panel resists bending because no ply can move independently of the others. It is precisely this coupling that the creasing rule must break locally along the fold axis so that the panel can hinge without tearing.

Cannella and Dai (2006) [20] quantified the effect over the full fold cycle. As the folding angle increases and the plies delaminate under the crease, the local bending stiffness of the corrugated structure falls by several orders of magnitude relative to the flat reference. This is not a marginal reduction. It is a controlled structural failure at a predefined line. The zone is only a few millimetres wide. It converts a stiff panel into a flexible hinge. Without it, no carton could be erected from corrugated board. The crease is a designed fracture.

Garbowski et al. (2021) [4] add the first measurable step of the same process. The creasing operation alone, before the first fold begins, already removes a significant share of the local bending stiffness through fibre deformation, local flute collapse and partial ply separation. Industry guidance formalises the trade-off as the fold factor: insufficient creasing leaves the fold resistance too high, and the liner tears on the outside of the hinge; excessive creasing destroys the liner, cracks the fibre network and produces crack initiation points that seed failure adjacent to the manufacturer's joint. The creasing tool operates inside a narrow therapeutic window and crushes the board just enough.

Where does the unwanted crushing come from

The dominant source of unwanted crushing sits in the machines themselves. Garbowski et al. (2021) [4] documented systematically how crushing develops during converting, with the largest contributions from the guiding elements in the corrugator and from the rollers and belts of the flexo folder-gluer. Jamsari (2020) [8] added laser-cut cross sections that show permanent flute deformation even at compression levels that leave the board looking visually intact.

At the corrugator, several components contact the freshly made board: the pressure rolls, the double backer, the slitter-scorer, the rotary shear and the stacker belts. The starch bond is still maturing. Any pressure beyond the flute's capacity causes permanent deformation. At the flexo folder-gluer, the vacuum belts, transport rolls, creasing wheels and folding belts each give a pinch. Short nips before the folding station are especially exposed: the flutes there still have the initial stiffness of the uncrushed board.

A third source is the strapping band holding the finished bundle together. A band tensioned at the delivery stacker presses the top and bottom flats against the inner population. The crush is local when the band sits across a flute trough; if it sits across a flute tip, it can propagate through several flats. A four-band strapping pattern that crosses at the corners leaves crushed zones that can be identified on the finished bundle with a side-light torch.

The pallet and the transport each contribute, but in most lines, they rank below machine-induced crushing. A forklift fork slightly off centre, a bundle leaning against a wall overnight, a corner pushed against the next pallet during transport: each leaves a zone of crushed flutes, usually local to one corner, with a pattern that differs from the distributed crushing left by the machines.

What changes when the blank arrives pre-crushed

The creasing tool does not choose the amount of damage it creates. It imposes a fixed geometry: a male rule, a female groove, a specified nip and a set penetration depth. The tool's effect depends on what it finds when the blank arrives. The setting assumes an undamaged reference: flute height at nominal value, symmetric flute flanks, intact ply bonds at the tips. When that geometry holds, the scatter is bounded, and the tool setting has a meaningful operating point.

Every reference cited for the desired damage presupposes this starting condition. The stiffness collapse reported by Cannella and Dai is measured on an intact board. The stiffness loss reported by Garbowski is measured on an intact board. The industry fold factor is specified for an intact board. None of these numbers describes what happens when the board, having already been crushed somewhere upstream, arrives at the creasing unit.

A blank that has passed through pressure rolls, folder-gluer vacuum belts or strapping bands no longer matches the reference state. The flute height is reduced. The flanks are weakened asymmetrically, as Jamsari (2020) [8] showed with his similarity factor for flute profile damage. Ply bonds at the tips may partly detach because the adhesive zone has absorbed local shear during crushing. The creasing tool now works on a substrate whose hinge behaviour depends on the blank's damage history, not on the tool's engineering.

Three mechanisms lose predictability at once. First, the extent of controlled delamination under the rule depends on the prior delamination already present. If the hinge forms earlier, or is laterally displaced, or if delamination is incomplete, residual coupling remains. The tool is then expected to have broken what was already half-broken; the result is not reproducible. Second, a weakened flute under the creasing rule does not compress along the rule axis. It tips sideways into whichever flank is weaker. The rule no longer strikes the flute at the tip, the shoulder or the trough, as defined by the geometry the operator set up. It strikes the tipped flute at whatever position the damage direction has chosen. Nagasawa's three contact cases collapse into a fourth one, the literature did not parameterise: the moving target. Third, the industry fold factor loses its window. Insufficient and excessive creasing are defined relative to the intact calliper. On a pre-crushed blank, the calliper is already reduced. The same rule penetration now sits deeper into the remaining fibre mass than intended. A setting that produces a clean crease on the calibrated reference overdrives the liner on the crushed blank, initiating cracks adjacent to the manufacturer's joint. The same setting on a less crushed blank later in the run may be correct again. The operator has no external signal to distinguish between the two cases.

The practical picture across the line

The practitioner sees the consequences at four observable places on the line.

1. **The crease edge.** A run of nominally identical blanks produces crease lines that differ in cleanness, sometimes on the same pallet. One blank shows a sharp, controlled hinge. The next shows a frayed edge where the liner has cracked. A third shows a double crease because the rule has bounced against a pre-delaminated zone and has not transferred its force cleanly.

2. **The fold line position.** A clean fold follows the crease axis. A fold on a pre-crushed blank drifts towards the weaker side of the tipped flute, because that is where the hinge of least resistance now lies. The fold line is no longer set by the tool. It is set by the damage geometry. The gap at the manufacturer's joint inherits the drift.

3. The spring-back force. Moistening the crease – so-called wet creasing – reduces the spring-back moment because the bending stiffness at the crease falls with the moisture dose. On a pre-crushed blank, the baseline spring-back is already variable because the baseline stiffness is already variable. The dose tuned for the reference calliper lands on a thinner, pre-weakened substrate that absorbs fluid faster and reaches saturation sooner. The therapeutic index narrows at both ends at once.

4. The manufacturer's joint itself. A blank with an unpredictable fold line presents the flap at a variable angle to the opposing liner. The adhesive bead meets an angled surface rather than a square one. Fishtailing appears at values the fold tool alone cannot explain, and the operator goes looking for a folding-bar misalignment that is not there. The cause sits upstream, in a crushing event that the creasing unit inherited rather than created.

Deflection under non-uniform stability

A bundle with uneven flute stability does not yield evenly. It gives way where the flutes have already surrendered and transfers the load to the areas that still stand. The whole bundle leans, the flats slide, the creases along the lean line open up. Flats that arrive at the case erector from a deflected bundle carry a mechanical pre-history that the operator cannot read by eye. The same principle acts on a smaller scale inside a single flat: a zone where the flutes were crushed at the folder-gluer nip is softer than the zone next to it. When the flat is fed through the case erector, the folding rails encounter differing resistance along the crease, the fold angle varies, and the fishtail returns even though the creasing geometry was correct. Garbowski et al. (2021) [4] quantified the loss and reported a steep non-linearity in which bending and torsional stiffness fall disproportionately once residual crushing exceeds moderate levels, with torsional stiffness collapsing earlier than bending stiffness.

7.4 Reading the Evidence on the Bundle

The corner slit is a window.

A bundle of knocked-down, glued flats keeps a record of the variation that produced it. The most honest window into that record is the corner slit, the relief cut at the corner where the manufacturer's joint meets the top and bottom flaps. Look along the cut edge of the slit across the side of a full bundle, and the evidence appears without any instrument at all. On one flat, the flute sits complete at the slit. Next, the flute is cut halfway through. On a third, the slit ends in the middle of a trough. On a fourth, it opens onto a full flute tip.

The pattern is more than visual interest; it is the direct fingerprint of the cut position relative to the flute wave, repeated one flat after another. Each flat captures the flute position of its moment of birth at the corrugator. When viewed side by side in the bundle, the flats write the drift of the upstream cut in plain sight.

What the pattern tells the practitioner

A repeating pattern along the bundle, where every slit shows a similar flute position, is evidence of a cut register that drifts slowly and within narrow bounds. Maintain the current settings: the upstream machine is holding its phase reference, and the creasing station has a predictable starting point.

A wandering pattern, where the flute position drifts from complete to cut to complete across the bundle, is evidence of register drift in the upstream cutting. Communicate the drift to the corrugator team for adjustment. What looks like a creasing problem at the converter is in fact a cutting drift at the corrugator, written clearly on the side of every bundle.

An erratic pattern, where no two slits look alike, and the flute position jumps from one flat to the next, is evidence of a mechanical disturbance: a worn knife, a slipping feed, a vibration in the rotary shear. The slit records the disturbance with millimetre accuracy.

Using the bundle as a diagnostic tool

This diagnostic costs nothing in hardware. It requires no sensor, no software, no calibration. A shift leader who walks past the finished bundles before they leave the converter and scans the corner slits along each bundle's side sees the quality of the upstream process in a single sweep. A drift shows itself without any need to interrupt the line.

The practice is worth writing into the daily routine. A quick scan before strapping, a photograph of any drifting bundle, and a note passed back to the corrugator with the pallet number should be standard. Over weeks, the pattern of patterns builds a picture of the upstream process that no single inline measurement can give. The bundle becomes an inspection sheet written in the product's language.

The caveat is consequential and must be stated openly. The method works only if the shift leader is trained to read the slits – to distinguish a repeating pattern from a wandering one, a wandering one from an erratic one, and all three from the merely unfamiliar. Training is the real investment, not the tool. A plant with an optical register monitor at the infeed but without a reading culture at the delivery stacker will see the skew before it happens and miss the cut drift for weeks. A plant with a trained shift leader at the bundle and no inline sensor at the infeed will reliably catch the cut drift, but will only see the skew after the bundle is finished. The two diagnostics are complementary, and the binding constraint for both is the same: someone has to know what they are looking at.

7.5 Fishtailing Revisited

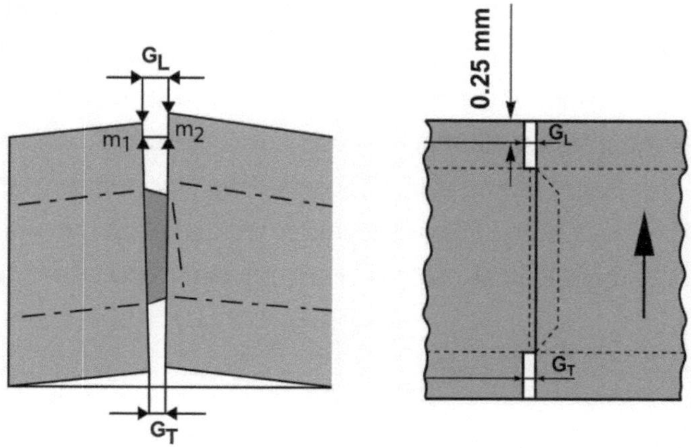

Figure 27 The two distinct geometric defects at the manufacturer's joint. Fishtailing (left): G_L ≠ G_T, the joint splays along the pack height. Parallel gap deviation (right): G_L = G_T but both off-spec, the whole joint shifted relative to the 0.25 mm tolerance band.

The mechanism in the literature

Fishtailing is the lack of parallelism at the manufacturer's joint, evidenced by the splay of the two edges when the erected box is viewed from above. FEFCO measures it as the difference between the front and rear gaps at the joint, with working tolerances set by the standard case style specification.

The primary cause, as documented in Nagasawa and Komiyama (2012) [29], Nica et al. (2020) [30], Wade et al. (2023) [18] and Jamsari (2020) [8], is a crease line that meanders rather than running straight. The meander arises when the creasing tool crosses the flute lay at an angle. At every transition from tip to flank to trough, the flute's lateral resistance changes, and the tool follows the path of least resistance.

Three secondary mechanisms amplify the primary cause. Washboarding displaces the adhesive asymmetrically across the joint, so that one side of the joint carries a weaker bond than the other and yields more under spring-back (Netz 1998 [11]; Johnson and Popil 2015 [9]). The warp energy stored in the flat from the corrugator is released at the fold, and an asymmetric warp pattern drives the splay further open on one side (Beck and Fischerauer 2022 [1]). Any pre-existing crushing from the converting nips reduces the bending stiffness locally and allows the fold to yield more at the weakened spot, an effect Garbowski et al. (2021) [4] directly tied to the converting process.

Parallel offset versus skewed fold

The two-fold signatures introduced in Section 1.4 map directly onto the literature. A fold that sits parallel-offset, with both edges of the joint running parallel but displaced from the target line, is the fingerprint of a straight crease at the wrong flute phase. The flute phase was wrong because the cut register drifted, because the infeed was offset, or because the whole blank sat off-centre in the creasing station. The crease line did not meander. It simply started from the wrong position along the flute wave.

The mechanism, as described by Nagasawa and Komiyama, is that the crease deflection differs substantially depending on whether the tool strikes a flute tip or a trough, resulting in a parallel deviation of the gap at the joint.

A fold that is skewed, with the two edges of the joint splaying outward, is the fingerprint of a crease that crossed the flute at an angle. Along the length of the crease, the tool encountered varying resistance and meandered. When the panels were erected, the fold angle varied along the crease, producing splay. The mechanism lies in the flute anisotropy described by Wade et al. and Jamsari, compounded by the washboarding-driven asymmetric glue bond and the warp spring-back.

The diagnostic consequence

The two signatures demand different responses, and the shift leader who confuses them chases the wrong cause. A parallel offset that runs consistently across many bundles is a call to the corrugator: check the cut register, the splice position and the knife condition. A parallel offset that appears only on some bundles is a call to the folder-gluer infeed: check the side guide, the vacuum table and the feed belt – or, if installed, read the inline register monitor directly rather than wait for the bundle to declare itself.

A fishtail that runs consistently across many bundles is evidence of a systematic skew, either at the corrugator slitter-scorer, at the folder-gluer creasing section or as the combined effect of warp and washboarding. A fishtail that appears intermittently points to the crushing pattern left by the strapping or by a worn nip in the folder-gluer, because the bending resistance changes locally.

7.6 Gap Deviations Revisited

The three gap modes

The gap at the manufacturer's joint varies across a batch in three distinct modes, each tied to a different chain of causes. The first mode is the uniform offset, where the gap is either too wide or too narrow but remains parallel and constant across the box height. This mode corresponds to the parallel-offset crease and arises from the flute phase at the crease line. Where the tool strikes a flute tip, the crease deflects deeper, and the fold sits closer to the target. Where the tool strikes a trough, the crease deflects less and the fold sits further out. The gap dimension shifts uniformly along the fold.

The second mode is the tapered gap, where one end of the joint shows a wider gap than the other, with parallel edges, but the joint visibly opens from front to rear of the box. This mode corresponds to the skewed crease – the fishtail signature – and points to a creasing axis that is no longer parallel to the flute axis.

The third mode is the erratic gap, in which the gap varies irregularly along the box height. This mode points to a local disturbance, most often a zone of crushed flutes left by a folder-gluer nip or a strapping band, or to a secondary fracture line next to the primary crease, as reported in the literature to be tied to asymmetric warp release.

What each mode does to the box

A uniform, narrow gap prevents the flaps from seating. The adhesive bead is squeezed, the joint carries stress, and the corner shows a raised line. A uniform, wide gap leaves a visible slot, misses part of the bonding surface and reduces the corner stacking strength. A tapered gap – the fishtail – produces uneven load distribution in the stacked pallet and jams the case erector because the guides no longer register a consistent outline. An erratic gap first appears during quality inspection, where boxes from the same order display different dimensions for no reason that the operator can immediately trace.

Systematic versus random deviations

The deepest practical distinction sits underneath the three modes. A gap deviation can be systematic or random, and the distinction decides whether the deviation can be corrected at the machine at all. A systematic deviation is consistent over time. It shifts the gap in the same direction and by a similar amount across a whole run, or in a predictable pattern that repeats from one bundle to the next. A random deviation jumps from one flat to the next without a pattern: wider on one flat, narrower on the next, centred on the third, without any traceable signature.

The systematic deviation is correctable. A gap that is consistently two millimetres too wide across a running order can be chased at the folder-gluer by shifting the folding bar inward. A gap that is consistently tapered with the same lean across all boxes can be corrected by adjusting the parallelism of the folding rails relative to the creasing axis. The corrugator can be called to check the cut register if the parallel offset points upstream. A shift leader who sees a stable pattern has a target to aim at. Each correction moves the gap in a known direction, and the outcome can be measured within minutes on the next few boxes.

The random deviation cannot be corrected at the creasing station or at the folding bar. It sits inside the spread of the substrate itself: a variable moisture signature across the incoming pallet, a variable crushing pattern from the corrugator nips, a variable flute phase from a drifting cut register, a variable skew from an unstable infeed belt. The folder-gluer can only work on the mean, not on the scatter around it. A shift leader who tries to correct a random deviation at the folding bar chases the last box and misses the next one, because the cause sat upstream and the next flat carries a different signature into the same machine.

The lever for the random part of the deviation is different. It is not a correction at the folder-gluer. It is a narrowing of the upstream variation: tighter cut register at the corrugator, equalised moisture across the warehouse, shorter residence times between process steps, gentler nip settings that leave less crushing behind, and cleaner infeed alignment that reduces the skew contribution. Each intervention narrows the spread that arrives at the folder-gluer, and the random deviation shrinks with it. What remains after all these upstream actions is the residual scatter that no line can eliminate – because variation is the normal state of the process, and removing it altogether was never on the table.

The diagnostic chain

The operator who reads a gap deviation has three upstream places to look, in this order. First, the flute phase at the crease, which tells the uniform-offset story. This is read directly from the corner slit on the bundle. Second, the creasing axis angle relative to the flute axis, which tells the tapered-gap story. This is read from a comparison of the front and rear gaps across a sample of boxes. Third, the crushing pattern from the folder-gluer nips or from the strapping, which tells the erratic-gap story. This is read from a side-light inspection of the bundle.

Before any of these three checks, the shift leader should answer a fourth question, ahead of them all: Is the deviation systematic or random? A systematic deviation justifies a correction at the folding bar, the creasing pressure or the creasing wheel. A random deviation redirects attention upstream to the pallet, the warehouse climate and the corrugator register. A shift leader who changes the creasing pressure in response to random scatter chases the symptom, adding a new source of systematic offset on top of the existing noise.

The order matters. A shift leader who changes the creasing pressure before checking whether the deviation is systematic treats the symptom and leaves the cause. A shift leader who reads the corner slit first, then the taper pattern, then the crushing signature, and who asks at every step whether the pattern is consistent or not, will find the cause in almost every case before touching a creasing tool setting.

7.7 Principal Variation Drivers and Their Controllability

Section 6.3 listed the seven principal variation drivers in tabular form, together with their causes and controllability. The sections of the present chapter have walked each of those drivers back to its mechanism: the flute phase position to the cut register and the lateral infeed alignment (Section

7.1); the washboarding amplitude to the starch add-on at the corrugator and the liner moisture (implied throughout 7.1 and 7.2, developed in Chapter 2); residual flute crushing to the belt pressures, fluting moisture and flute anisotropy of the converting line (Section 7.3); warp to the differential stresses and storage climate of the pallet (Section 7.2); liner porosity to the batch and moisture history of the feedstock (Section 7.2); adhesive viscosity to the temperature, shear and pot life of the application unit (Chapter 3); and pallet position to the deck-board gaps, overhang and stacking pattern beyond the manufacturer's gate (Chapter 4).

The summary table that ties these threads together appears in Section 6.3 and is not repeated here. The important point, once the mechanisms have been laid out row by row, is the pattern of the right-hand column: most of the drivers are not controllable, or only partially so. The flute phase position is physically unavoidable. Liner porosity varies from batch to batch. The position of a box on the pallet is beyond the manufacturer's reach altogether. Conventional parameter optimisation can address only controllable factors, which are in the minority.

The craft lies in shifting the weight of effort from tweaking parameters at the folder-gluer to narrowing the upstream spread – and, where upstream narrowing has reached its limit, to building robustness into the downstream structure so that the remaining scatter no longer translates into failure. The first of these moves is the subject of the diagnostic practice developed in this chapter. The second is the subject of Part IV: stitching for damage tolerance at the joint (Chapter 10) and waterscoring for local conditioning at the crease (Chapter 11).

Chapter 8:
The Secondary Fracture Next to the Crease

Why the Hinge Sometimes Forms Where the Tool Did Not Draw It

Introduction card

A fine crack beside the score line is not a cosmetic defect. It is a diagnostic signal. This chapter explains how hairline cracks, parallel grooves, and frayed zones reveal whether the cause lies in the tool, flute phase, moisture, pre-crushing, warp release, or false scoring. It is valuable for troubleshooting, supplier development, SPC logging, and print-quality investigations.

"*What produces the parallel hairline crack next to the score line, and what does its appearance tell the practitioner about which station upstream needs to be corrected?*"

Quality: Six causes produce three main visible signatures: a hairline crack, a parallel groove, or a frayed zone. The causes are crease pressure outside the therapeutic window, flute phase at the rule, moisture-induced fibre brittleness, pre-crushed or pre-delaminated zones, warp-release loading, and false-score competition from an adjacent flute. The cross-table in the chapter links each cause to each signature in both directions. The chapter also adds the IGF 21804 BR finding for parallel scores: wave-crest creasing imposes the lowest stress on the upper liner, valley creasing imposes the highest, and intermediate-position creasing produces an asymmetric crease bead that is geometrically unrecoverable. Parallel creases should therefore be performed as wave-crest creases by maintaining a constant longitudinal positioning of the corrugated board sheet relative to the rule.

Operational efficiency: Each visible signature points to a different corrective action. Tool pressure is adjusted at the rule. Flute phase cannot be adjusted and must be designed around. Moisture equalisation belongs in the warehouse. Warp management belongs at the corrugator. Pre-crushing must be traced to conveying or transfer elements upstream. The chapter therefore acts as a routing map for the audit.

Economic impact: Linear fractures at the fold edge weaken the area next to the manufacturer's joint. Adhesive selection cannot compensate for this damage. On digitally printed corrugated, the related print-side defect appears as a fine white crack along the folded print edge. This can affect premium-print orders even when the buyer cannot immediately name the technical cause.

Regulatory and quality systems: The extended defect-code system gives plants a structured vocabulary for SPC, supplier complaints, and incoming-goods inspection. It records where the crack occurs, how far it runs, and which side of the board it affects.

Limitations: Real defects often combine several causes. The diagnostic table identifies the dominant mechanism, not every contributing factor. The correction often belongs upstream of the rule, even when the visible defect appears at the crease.

Read in full: production engineers, quality managers, supplier-development teams.

Card sufficient: executives reviewing the substrate-side defect framework; print buyers diagnosing print-side fold-edge defects.

Prerequisite: Chapter 7.

Follow-up: Chapter 11, Mechanism 3, waterscoring as the diagnostic-loop closure.

A creasing tool is a precision instrument that draws a single line. On some blanks, the operator sees two. Next to the intended crease, a parallel fracture has opened in the outer liner — a thinner but unmistakable second groove, offset by a few millimetres, sometimes running parallel for the whole fold length, sometimes only part of it. The industry calls this phenomenon the Nebenbruch, the secondary crease, or the false score.

The fold no longer opens cleanly at the designed axis. It drifts towards the secondary fracture, and the geometry at the manufacturer's joint follows this drift. Its causes are several, and the literature has mapped them one family at a time.

The secondary fracture does not sit beside the variation picture by accident. It is one of the most concrete signatures through which the random part of variation prints itself on the finished blank. Where Chapter 7 described the spread of the fold and its upstream origins, this chapter reads the trace the spread leaves next to the rule.

8.1 The Liner Is a Paperboard

Most of the detailed fracture mechanics of the creasing process were worked out on solid paperboard rather than on corrugated board. That is a matter of convenience, not of physics. The outer liner of a corrugated panel is itself a paperboard sheet, manufactured on a Fourdrinier or a multi-ply former, in the same way as the cartonboards tested in the cited literature.

The tensile strength of the liner at the groove shoulder, its fracture toughness under tension, its dependence on fibre orientation and moisture, and the shear zones that form to either side of the rule are properties of the liner sheet alone. None of them changes when the liner is subsequently glued onto a fluted medium.

The corrugated case adds three elements to the mechanics of paperboard. The flute introduces a cyclically varying support that alters the stress distribution in the liner along the crease — a flute-phase effect. The bond to the flute tip introduces a delamination degree of freedom that flat paperboard lacks, enabling controlled crushing under the rule. The two liners, coupled through the medium, introduce a composite stiffness in the flat panel that collapses dramatically as the crease forms, as reported by Cannella and Dai [20].

The mechanisms of the secondary fracture sit in the liner layer that the paperboard studies measured. The shoulder crack, the flute-tip shift, the moisture-related brittleness, the microcrack at an overdriven rule, the broadened crush band at a plasticised rule — all of these are liner events. The corrugated case modulates them through the flute phase, the pre-crushing history of the flutes, and the adhesive bonds, but it does not replace the underlying fracture mechanics. Reading Beex and Peerlings [98] or Huang, Hagman and Nygårds [100] on paperboard is therefore reading the mechanics of the liner in a corrugated panel, with the flute subtracted. Adding the flute back brings Nagasawa and Komiyama [29], Nica et al. [30], Wade [18], Jamsari [8] and Garbowski [4] into the picture.

8.2 What the Practitioner Sees

The secondary fracture appears in three forms.

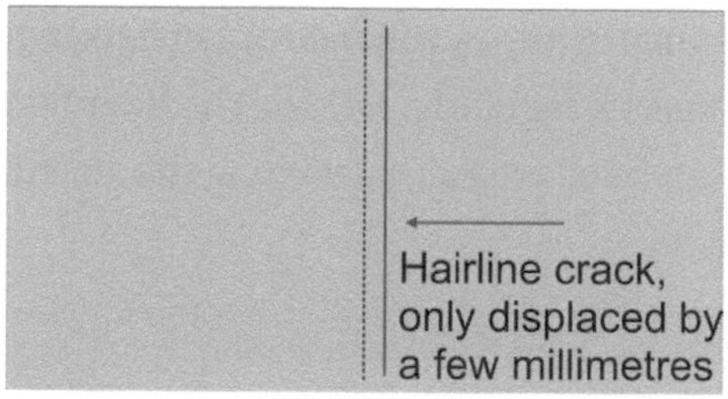

Figure 28 Hairline crack — a faint parallel line a few millimetres beside the rule, visible only under grazing light

First, there is a hairline crack in the outer liner, a few millimetres from the tool axis, visible only under grazing light on an unprinted blank.

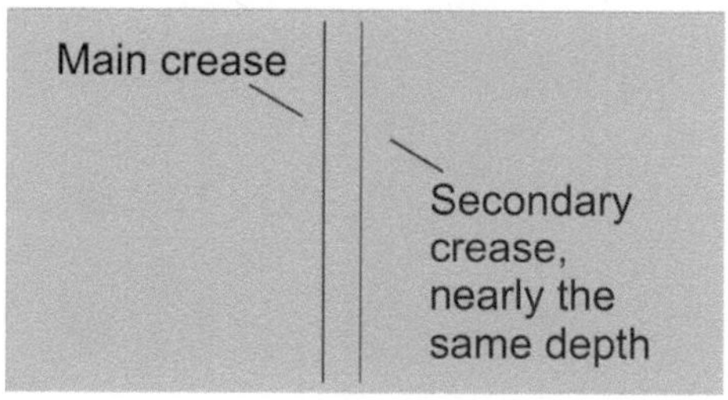

Figure 29 Parallel groove — a second crease nearly as deep as the primary.

Second, a parallel groove appears, nearly as deep as the primary crease, making the blank look as though it has two creases where the tool drew one.

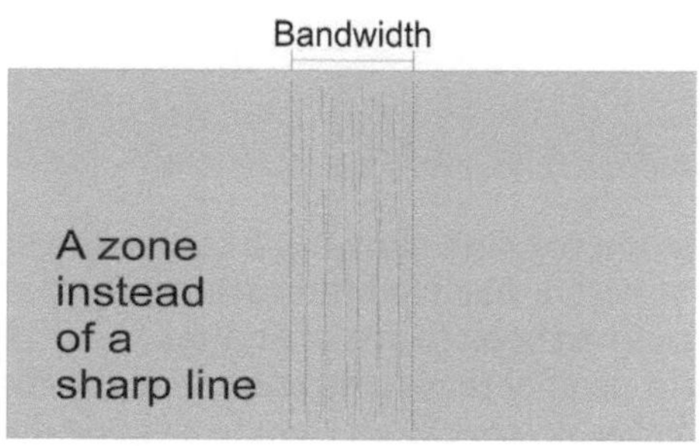

Figure 30 Frayed zone — a widened crush band where fibres have pulled apart irregularly.

Third, a frayed zone may form—a widened crush band where the fibre network has pulled apart irregularly, leaving the liner feathered. All three arise alongside the intended crease, weakening the substrate in the otherwise intact region.

Figure 31 Real crease photograph showing a composite signature — a broad frayed zone (Form 3) with embedded indications of a parallel groove (Form 2). In practice the three textbook forms of Figures above rarely appear in pure isolation; the operator usually reads a mixture, and the cross-table in Section 8.10 must be applied row by row rather than to a single dominant signature.

The three forms above are a teaching reduction. For systematic logging —

quality-control records, supplier complaints, and the buildup of a per-line defect history — the finer ordinal scheme of the VDW Prüfkatalog, as extended in IGF 21804 BR [103], is more useful. The codes below distinguish where the crack sits (inside the bead, beside it, or perpendicular to the rule), how far it runs (continuous or partial) and which side of the board carries it. The Fehleranteil parameter at the bottom of the table converts a per-specimen yes/no log into a population-level metric suitable for SPC tracking of a creasing line.

Code	Description	Side
0	No crack	either
2a	Outer-liner crack inside the bead — continuous over the full crease length	outer
2b	Outer-liner crack inside the bead — partial	outer
2c	Outer-liner crack alongside the bead — continuous over the full crease length	outer
2d	Outer-liner crack alongside the bead — partial	outer
3a / 3b	Crack perpendicular to the crease line — continuous (3a) or partial (3b)	transverse
4a	Position of the bending line is displaced from the rule axis	either

Code	Description	Side
5a	Outer-liner microcracks (the classic hairline of Section 8.2)	outer
7a / 7b	Outer liner cut by the rule — partial (7a) or full-thickness (7b). Failure of the therapeutic window of Section 8.3.	outer
1a	Inner-liner crack inside the groove — continuous	inner
1b	Inner-liner crack inside the groove — partial	inner
1c	Inner-liner crack beside the groove — continuous	inner
1d	Inner-liner crack beside the groove — partial	inner
6a / 6b	Inner liner cut by the counter-die — partial (6a) or full-thickness (6b)	inner
8a	Folds on the inner side (the frayed-zone signature of Section 8.2)	inner
FA	Fehleranteil = (number of defective specimens / sample size) × 100 %. Population-level summary metric.	either

Mapping back to the three forms of Section 8.2 and the cross-table of Section 8.10: the hairline crack corresponds principally to codes 5a or 1d; the parallel groove corresponds to 2a/2b on the outer side and 1a/1b on the inner side; the frayed or wandering crush band corresponds to 4a (drift of bending position) plus 8a (inner folds), often co-occurring. The cross-table in Section 8.10 reads identically, with the codes substituted for the form labels; the codes simply make the row-by-row diagnostic more fine-grained.

Where the primary crease is the hinge, the secondary fracture forms a parallel hinge. The bending moment from the folding bar no longer moves only through the tool axis. It follows the path of least resistance, which may be the primary crease on one blank and the secondary fracture on another. The fishtail and gap deviation at the manufacturer's joint follow directly. The mapping of these three signatures back to their six underlying causes is summarised in the cross-table at the end of this chapter (Section 8.10).

8.3 The Therapeutic Window of Crease Pressure

The first and most widely reported cause lies in the tool setting itself. Industry guidance and trade handbooks formalise the fold factor as a controlled range for crease depth and strength—a narrow band between two failure modes. A shallow crease leaves fold resistance too high.

The bending load during folding is carried by the outer liner in pure tension, and the liner cracks under strain. The crack does not necessarily form on the tool axis. It forms wherever tensile stress first exceeds fibre strength. In a non-uniform substrate, where material properties vary across the sheet, this can be a few millimetres to either side of the intended hinge.

Beex and Peerlings [98] and Huang et al. [100] documented this transfer of the fracture site away from the tool axis when the hinge is weakened, on paperboard specimens whose mechanics translate directly into the liner of a corrugated panel.

A deep crease inverts the failure mode. The rule penetrates far into the fibre mass, overstretching the liner at the groove shoulders, and crack-initiation points form on both sides of the rule. The primary crease now carries a pair of parallel microcracks that open fully during folding. Nagasawa et al. [101] showed, by finite element analysis, that creasing stresses develop not only under the rule but also in nearby regions, and multiple delaminations initiate in the shear zones on either side of the tool. The overdriven rule plants the seeds of its own secondary fracture.

The working window between these two failure modes is narrow. Industry guidance from C and T Matrix [99] specifies a score width in the range of roughly one and a half to two and a half times the board calliper, and a bead height in the range of roughly half the calliper. Outside this band, one of the two parallel cracks becomes almost unavoidable. The window is even narrower on double-wall and lightweight liners, where calliper tolerance and fibre strength each reduce the margin.

This is an industry rule of thumb, not a formal specification. The DIN 19304:2021-04 standard for corrugated-board crease testing prescribes how to measure the fold factor. However, it does not define a target range. The IGF 21804 BR investigation [103] confirmed that no validated optimum range for the corrugated-board fold factor exists in published literature. The DIN 19304 (for corrugated board) and DIN 55437-3 (for cartonboard) standards also differ in terminology, key figures, and quality-assessment scales. This can confuse practitioners who work with both substrates. The IGF authors raised this discrepancy with the responsible DIN committee NA 074-02-03 AA. Thus, the therapeutic window is defined by practitioners and tool suppliers, not by a standard. This makes it weaker as an audit reference, but not less binding in practice. The variation argument in Chapter 6 is only sharper when the window itself is a community estimate.

8.4 Flute Phase and the Shoulder as a Shear Maximum

The second cause is geometric. Nagasawa and Komiyama [29] showed how the position of the creasing tool relative to the flute tip determines where the liner is most strained. When the tool strikes a flute tip, the liner is squeezed between the rule above and the flute tip below.

Tensile stress in the liner peaks not directly beneath the rule, but at the shoulders. Here, the liner bends around the tip and stretches the most. The liner then fractures at the shoulder. The resulting crack runs parallel to the rule but is offset by the distance between the tip and the shoulder. This creates a secondary fracture of geometric origin.

Nica et al. [30] confirmed this picture for inline creasing at the corrugator. Because the crease is applied in the machine direction, the phase between the tool and the flute wanders continuously. This suggests that maintaining a consistent phase or adjusting tool timing can reduce the variability in fracture timing. A tip contact and the paired shoulder fracture alternate with a trough contact and a different, gentler stress field.

On the same blank, along a single fold, the secondary fracture appears in some portions but not in others. The operator sees a crease that cracks only where the tool meets a flute tip.

The sign of the flute-position effect depends on the orientation of the crease.

The Nagasawa–Komiyama mechanism described above is specific to creases running transverse to the flute direction. Along such a crease, the tool sweeps across many flutes in succession. It alternates between tip contacts , where the shoulder shear maximum forms and a hairline initiates, and trough contacts, where the stress field is gentler. In this geometry, tip contact is the adverse condition.

For creases parallel to the flute direction, the entire crease lies along a single position relative to the wave geometry. The geometry has three distinct cases: wave-crest (Bergrillung), valley (Talrillung), and intermediate (Zwischenrillung). The IGF 21804 BR finite-element investigation by Harling and Schlimpert [103] paints a different picture. For parallel creases, tip contact is the gentlest condition.

Their LS-DYNA simulation was validated against laboratory-press samples using a Marbach test die. It tracks the MD-stress in the upper liner directly under the rule across the rule's downward stroke. For a wave-crest position, the same linear element shows the lowest initial rise and the lowest global maximum. In a valley position, it experiences both a steeper initial rise and a higher peak stress. The intermediate position, midway up the wave flank, produces an inherently asymmetric crease bead. One flank is supported under the rule; the other is unsupported. This is for the same mechanical reason that a pre-crushed flank produces an asymmetric hinge in Section 8.6.

The implication for production is direct: perform parallel creases as wave-crest creases. This is done by maintaining a constant longitudinal positioning of the corrugated board sheet relative to the rule. The valley case is mechanically the worst. The intermediate case is unrecoverable because the asymmetry is geometric rather than statistical. The same flute-tip contact that the transverse-crease operator wants to avoid is the contact the parallel-crease operator wants to guarantee. Tip contact is not universally adverse — its sign is set by the crease orientation.

8.5 Moisture and Fibre Brittleness

The third cause lies in the board's material condition. A single material property — the elongation at break of the outer liner — ties the others together. The IGF 21804 BR study [103] correlated 159 measured material parameters with crease quality across 11 corrugated grades and 3 climate steps. It found no single material parameter that dominated the diagnosis. The strongest individual correlations (ECT, SCT, FCT, burst strength, PAT, S-test) ranged from 0.3 to 0.47. The qualitative finding from their failure-case analysis was sharper than any single coefficient. Papers with high elongation at break consistently produced stable creasing processes. Papers with low elongation produced unstable ones.

Moisture, the variable that has dominated the historical literature on this question, is best read as one input into elongation rather than the underlying mechanism. Cornaggia et al. [3] and Szewczyk and Głowacki [15] documented the moisture dependence of linear tensile strength and fracture toughness directly. Below approximately 8% moisture, the fibres lose their elongation reserve and become brittle. The same brittleness can also arise from a kraft liner with intrinsically low elongation, or from a liner overstretched at the corrugator. All three pathways converge on the same secondary-fracture signature.

The crease that the tool draws on a calibrated, eight per cent substrate is the same crease that fractures into a secondary groove on a four per cent substrate. The primary crease still forms, but it cannot absorb the bending energy. The excess energy finds its way into a parallel crack in the drier, more brittle fibre population alongside the rule.

Moisture gradients across the pallet cause the issue in batches. To prevent this, acclimate pallets in a controlled environment before processing. When a pallet is left overnight in a heated warehouse, the dry outer sheets arrive at the folder-gluer with a fibre network near its brittle limit, while the core sheets do not. As a result, fifty blanks run through the same tool and settings — half develop a secondary fracture while half do not. The tool stayed constant; the substrate changed.

At the opposite extreme, when moisture exceeds roughly 14%, the fibres become too plastic for a defined crease. The tool prints a broad crush band rather than a line, and the folder shows no crisp hinge under load. The appearance is less a secondary fracture than a blurred primary one. The underlying mechanism remains the same: the substrate has not absorbed energy as the tool geometry assumed.

If low elongation at break predominates, whatever its origin, the diagnosis clearly points to a therapy discussed later in this book. Chapter 11.4, Mechanism 3 (Increased Liner Elasticity), explains how a controlled, local water dose at the crease line raises liner ductility by disrupting fibre-fibre hydrogen bonds. Softened fibres deform plastically rather than fracturing, preventing parallel hairline initiation. This cross-reference is meaningful. The IGF 21804 BR sensitivity analysis [103] identified elongation at break as the sole material-side parameter consistently linked to stable creasing. Waterscoring is the process that delivers elongation on demand. Diagnosis and treatment are clearly separated.

8.6 Pre-Crushed and Pre-Delaminated Zones

The fourth cause was described in Chapter 7 and is returned to here with greater clarity. Jamsari [8] demonstrated, using laser cross-sections, that a flute crushed upstream presents asymmetric flanks to the creasing tool. One flank is weaker than the other. The hinge under the rule is not symmetric and shifts towards the weaker flank. Meanwhile, the original flute tip is no longer a geometric peak; it has flattened. The rule now applies to a broader contact area with a stress distribution having two maxima, one on each flank. A secondary fracture can form on one flank while the primary fracture forms on the other.

Garbowski et al. [4] added the statistical picture. On the B-flute and C-flute boards, residual crushing beyond a moderate threshold collapses torsional stiffness so that the hinge response becomes unpredictable from blank to blank. Cannella and Dai [20] showed that ply delamination under the crease leads to a steep drop in stiffness at full fold. Where plies are already partially delaminated before the tool engages, the tool no longer initiates delamination. The delamination front starts earlier along the stroke, or laterally displaced from the rule axis, or both. The resulting hinge is not the single line the tool drew; it is a pair of lines, one under the rule and one along the pre-existing delamination front.

One source of pre-crushing overlooked in the historical literature is the rule itself, specifically regarding transverse creases. The IGF 21804 BR finite-element investigation [103] of corrugated board cross-creased perpendicular to the flute direction shows important details. During the rule's downward stroke, each flute flanks the rule kinks several times. The material then folds onto itself in multiple layers. This accumulated material is driven by the rule into the lower liner. As a result, it produces periodic plastic deformations throughout the sample, leaving permanent flute-height changes that persist after the rule retracts. Pre-crushing on transverse passes is therefore not only inherited from upstream — it is generated at the rule. A clearer reading of the section title is that any zone whose flute geometry has been disturbed before the current fold begins presents the same asymmetric flank to the next folding moment, regardless of whether the disturbance came from the corrugator, the conveying nip, the strapping band or the previous rule strike. The diagnosis from the bundle is the same, but the responsibility for the cause now points at one further station upstream — or, on transverse passes, at the rule itself.

8.7 Warp Release and the Third Axis of Loading

The fifth cause comes from the board itself, not the tool. A blank carries warp energy stored during the corrugator cooling curve, which is released during folding. Beck and Fischerauer [1] modelled how asymmetric warp patterns distribute spring-back force unevenly across a panel. On an asymmetrically warped blank, one side of the crease receives more spring-back moment than the other. This difference can drive a secondary fracture on the more heavily loaded side while the opposite side folds cleanly.

The phenomenon is particularly vivid on lightweight liners, where the warp spring-back force is comparable to the tensile strength reserve of the outer liner. Any local imbalance can cause the liner to transition from intact to fractured along a line parallel to the crease. Netz [11] and Johnson and Popil [9] linked the same mechanism to washboarding, where adhesive displacement across the flute wave produces local stiffness variations that amplify warp release along specific bands of the panel.

8.8 The False Score Across Adjacent Flutes

The sixth cause has its own name in industry terminology. Where the flute adjacent to the crease line is weaker than the crease line itself — whether because of localised crushing, a run of low grammage in the fluting, or a zone of poor flute-to-liner bonding — the blank folds around the adjacent flute rather than around the intended crease. C and T Matrix [99] describe the phenomenon as the 'false score' or 'rolling score'. The fold line rolls along the substrate, finding whichever flute currently offers the lowest resistance, and the hinge can wander by one or two flute pitches across the length of the fold. What the operator registers as a secondary fracture is, in fact, a competing hinge that the adjacent flute has offered to the folding bar.

The false score is distinct from the parallel fracture caused by tool overdrive. Its signature is not a clean hairline next to the rule but a broader, wandering band of buckled fibres that can be half a flute pitch wide. It appears more often on inline-creased boards than on offline-creased boards, because inline creasing offers the flute phase no protection, and the adjacent flute may be structurally indistinguishable from the target flute when the tool arrives.

8.9 Reading the Signature on the Blank

Each of the six causes leaves a slightly different signature next to the primary crease. The practitioner who reads them can separate tool faults from substrate faults without any instrument. A clean hairline, a few millimetres beside the rule, running the full crease length, points to tool pressure outside the therapeutic window.

A fracture that appears and disappears along the fold, alternating with clean sections, points to the flute phase: flute-tip contact produces the fracture, trough contact does not. A secondary fracture that runs through a whole population of blanks — rather than through individual points — points to a moisture gradient across the pallet. An asymmetric fracture, present on one side of the crease but not the other, points to warp release or washboarding. A broadened crush band displaced by a full flute pitch points to the false-score mechanism. A double line with neither line exactly under the rule points to pre-crushing in the upstream process.

None of these signatures can be solved by the creasing tool alone. Tool pressure can be adjusted. The flute phase cannot. Moisture equalisation belongs to the warehouse. Warp management belongs to the corrugator. Pre-crushing is part of the conveying elements upstream. The operator who reads the signature correctly directs the correction to the right place, instead of adding a tool change on top of a substrate problem.

8.10 Six Causes × Three Signatures: A Cross-Table

Sections 8.2 introduced three visible forms — hairline crack, parallel groove, frayed or wandering crush band — and Sections 8.3 to 8.8 traced six underlying causes. The two lists do not map one-to-one. A single cause can print more than one signature (the therapeutic window in Section 8.3 produces a hairline at the shallow end and a parallel groove at the deep end), and a single signature can be reached from more than one cause. The cross-table below makes the mapping explicit so that a blank in the operator's hand can be read in both directions: from signature back to candidate causes, and from a suspected cause forward to the signature that should appear if the diagnosis is correct.

8.3 Crease pressure outside the therapeutic window	Shallow rule: the liner cracks where tensile stress first exceeds fibre strength, a few millimetres beside the axis.	Deep rule: paired microcracks at both groove shoulders open into a second groove during folding.	—
8.4 Flute phase and shoulder shear	Tool meets a flute tip; bending around the tip stretches the liner at the shoulder. Appears intermittently along the same fold.	—	—
8.5 Moisture and fibre brittleness	Below ~8 % moisture: dry fibres fracture brittlely along a parallel line. Affects whole populations of blanks from the same dry pallet zone.	—	Above ~14 % moisture: fibres deform plastically; the rule prints a broad crush band rather than a defined line.
8.6 Pre-crushed and	—	Asymmetric flute flank shifts	—

pre-delaminated zones		the hinge; the primary forms on one flank, the secondary on the other. Neither line sits exactly under the rule.	
8.7 Warp release and the third axis of loading	Asymmetric spring-back fractures the more heavily loaded side of the crease; the opposite side folds cleanly.	—	—
8.8 False score across an adjacent flute	—	—	Fold rolls onto a weaker neighbouring flute; a broader, wandering buckle band, up to half a flute pitch wide, replaces the intended hinge.

Read row-wise, the table answers the question "if cause X is dominant, what should I expect to see?" Read column-wise, it answers the inverse: "a hairline crack appears on this batch — which causes are still in play, and which are ruled out?" A hairline that runs the full length of every blank in a population, for example, eliminates the flute-phase row (which would alternate along the fold) and the false-score row (which produces a band, not a line), and leaves Sections 8.3, 8.5 and 8.7 as the surviving candidates. The diagnostic step that follows is targeted rather than exploratory.

The dashes are the diagnostically informative cells. They mark the signatures that a given cause does not produce, and they are the reason the table earns its place: the absence of a signature is itself evidence. A pre-crushed-flank diagnosis (Section 8.6) that is paired with a frayed crush band, rather than a parallel groove, is almost certainly the wrong diagnosis.

8.11 A Thought to Close

The flute phase at the crease is not a parameter a technician sets once and leaves in place. It is the result of a chain of small inaccuracies, each one harmless on its own: the cut drift at the corrugator, the offset at the folder-gluer infeed, the skew of the blank across the creasing wheels, the moisture gradient between the core and the edge of the pallet, and the crushed zone left by a folder-gluer nip or by the strapping band. In combination, these inaccuracies decide whether the box sings or creaks when it is erected.

The evidence is written on the bundle itself. The corner slit along the side tells the story of the cut register without asking for a sensor. The fishtail at the joint tells the story of skew and washboarding in one glance. The drifting gap tells the story of the flute phase, the creasing axis and crushing all at once. A shift leader who learns to read the bundle as a document reads the variation of the entire line from one look at the strapped stack.

Crushing at the crease line is a precision tool. Crushing before the crease line is a rumour about what the precision tool will find when it arrives. The first is quantified, localised, and designed into the fold factor of the industry guidance and into the stiffness-collapse curve of Cannella and Dai [20]. The second has no design authority. It is the uninvited co-author of every crease the tool tries to make, and the reason a machine setting that worked on the first pallet can fail on the second.

The creasing tool draws one line. The blank answers with between one and three — and sometimes four, when the adjacent flute joins the conversation. Every secondary fracture is the board telling the tool that the therapeutic window was off, that the flute phase was wrong, that the fibre was too dry, that the flank was already crushed, that the warp was asymmetric, or that a neighbour was weaker. The crease is a designed fracture. The Nebenbruch is its uninvited sibling, and the mechanics of each of the two are known.

The liner of the corrugated panel is a paperboard. What the paperboard literature says about fibre brittleness, about crack initiation at the groove shoulder, about shear zones on either side of the rule, reads directly as a statement about the liner of a corrugated box. The flute on top modulates the stress but does not replace the mechanics underneath. A chapter that cites paperboard research to explain a corrugated fracture is not borrowing from a distant field — it is reading the corrugated liner through the instrument that measures it best.

Variation is not a flaw of a badly managed line. Variation is the normal state of every line. The craft lies in narrowing the band where it bites, in compensating for the part that remains, and in knowing which signature points to which upstream cause. The tool alone can never close that window. The line can — at every stage upstream of the rule — tighten its own contribution a little.

Diagnosis ends here. Chapter 8 lists the six causes and the three signatures, and shows how to walk between them in either direction. The therapeutic counter-moves belong to the chapters that follow. *Chapter 11.4 in particular — and within it, Mechanism 3 (Increased Liner Elasticity)* — converts the dry-fibre row of the cross-table into an actionable intervention: a controlled, local water dose at the crease line shifts the liner from brittle to ductile, and the parallel hairline that the diagnosis predicted simply does not initiate. The Nebenbruch does not have to be tolerated. It has to be understood first, and then engineered out, one row of the table at a time.

Chapter 9: Adhesive Application Defects: Splatter, Misplacement and Their Consequences

Introduction card

A tiny adhesive defect can stop a case erector. This chapter explains how splatter, satellites, tailing, bead misplacement, nozzle build-up, and viscosity drift create reopens, contamination, stoppages, and complaints. It is relevant for readers responsible for adhesive systems, downtime, sensor selection, or process control.

"How do satellites, splatter, tailing, bead misplacement, and viscosity drift at the folder-gluer become reopens at the case erector, and which defects can be eliminated by hardware rather than only monitored by SPC?"

Quality: The chapter defines six adhesive application defects and their consequences. Satellite formation and splatter can contaminate surfaces that must remain adhesive-free. Tailing is the filament before breakup and contributes to both satellites and nozzle build-up. Under-application can later appear as pop-opens. Over-application can cause strike-through and unwanted panel bonds. Dried adhesive after standstill can contaminate the first blanks after restart. Bead misplacement on narrow PPWR tabs can stiffen the crease zone and create erection problems.

Operational efficiency: A single misglued pack can stop the case erector and back up the upstream line. On poorly controlled lines, 10 to 20 incidents per shift can accumulate into substantial weekly downtime. Multi-band IR sensors can detect small splatter defects at high frequency, allowing the defect to be caught at source rather than at the case erector.

Economic impact: Adhesive viscosity can drift by roughly ±15 to 20 per cent across an unconditioned hall climate of 18 to 24 °C. A cold morning can produce high viscosity and tailing. A warm afternoon can produce low viscosity and needle chatter. The practical response is a valve specification matched to the actual hall climate, not full hall conditioning, which is usually far more expensive.

Regulatory context: Multi-band IR sensors do not require optical brighteners in the adhesive. This avoids food-contact restrictions linked to stilbene-class brighteners and prevents contrast problems on coated white liners.

Process: The chapter distinguishes defects that require hardware correction from those that can be reduced by settings, maintenance, adhesive specification, or monitoring. The aim is to prevent misapplication at the folder-gluer before it becomes a stoppage at the case erector.

Limitations: There is no commercial waterscoring sensor at the current state of the art. The SPC discussion applies to adhesive sensing. Waterscoring remains open-loop, with verification downstream through crease and bond inspection.

Read in full: adhesive specifiers, valve engineers, quality managers.

Card sufficient: procurement teams scoping a sensor or valve upgrade.

Prerequisite: Chapter 8.

Follow-up: Chapter 10, stitching as a damage-tolerance pattern.

The preceding chapters have shown how washboarding, warp, liner type and crease quality influence the bond at the manufacturer's joint. Yet even when all these parameters are within their optimal ranges, a single defect class can still bring the entire process to a standstill: faulty adhesive application. The signatures are familiar from the shop floor — splatter on supposedly adhesive-free surfaces, satellite droplets, misplacement of the glue bead relative to the crease line, and tailing threads between segments.

These defects originate in the application technology, but they propagate through material physics — viscoelastic thread formation, dispersion drying, and surface tension at the nozzle. They strike the value chain at a particularly vulnerable point: the case erector. This chapter analyses the defect patterns, their causes, their consequences and the technical solutions available today to prevent and monitor them.

A note on sources: The technical examples in this chapter are mainly taken from the published portfolio of one supplier (Baumer hhs), as I work there and have access to relevant data and photos.

9.1 Adhesive Application Defect Patterns

At the folder-gluer, six fundamentally different application defects can occur, each with distinct causes, signatures, and consequences along the downstream value chain.

Satellite formation and splatter

When a bead of adhesive exits the nozzle, a primary droplet forms and, depending on viscosity, machine speed and nozzle condition, one or more secondary droplets — known as satellites — form. These satellites take one of two paths: they travel uncontrolled and land on the substrate surface, often far from the intended glue bead, or they accumulate at the nozzle tip, accelerating nozzle build-up and causing them to be ejected as particles during subsequent switching cycles.

The physical cause lies in the viscoelasticity of the PVAc dispersion. When the valve closes, the adhesive strand is sheared, but the polymer solution draws a thread that breaks only after a typical length, then disintegrates into a primary droplet and one or several satellites. The higher the viscosity and the lower the valve's closing force, the longer the thread and the more satellites are generated during break-up. A low-performance valve with a slow closing time produces systematically more satellites than a valve with a high closing force and short switching time.

Tailing

In discontinuous application, where the valve opens and closes many times per second to deposit segmented glue beads, a thin adhesive filament forms between the closing nozzle and the substrate at the trailing end of each segment. This filament — known as tailing or stringing, and in the hot-melt vocabulary as angel hair or cobwebbing — does not break cleanly. It either lays down as a fine thread on the substrate beyond the intended segment end, draws back towards the nozzle and contributes to nozzle build-up, or breaks mid-air, releasing microdroplets that contaminate the substrate and surrounding machine components.

The mechanism is the viscoelastic response of the PVAc dispersion to rapid extensional flow. As the valve closes, the adhesive column between the nozzle and the substrate is stretched. A Newtonian fluid would separate quickly under gravity and surface tension; a polymer dispersion resists separation through chain entanglement and forms a filament whose break-up length scales with the dispersion's relaxation time. Higher viscosity, longer relaxation time and lower valve closing speed all increase the filament length. At typical folder-gluer speeds of 300 to 400 m/min, even a filament of one or two millimetres results in a measurable trailing deposit on the next segment, the next blank, or the machine surroundings.

Tailing is closely related to satellite formation but distinct from it. Satellites originate from the break-up of a stretched filament into multiple droplets; tailing is the filament itself, before break-up. Tailing is therefore the precursor defect from which both satellites and accelerated nozzle build-up emerge. This sequencing matters for inline monitoring: a system that detects only satellites or only nozzle deposits reacts to symptoms, whereas a system that detects the filament reacts to the cause.

Under- and over-application

The amount of applied adhesive determines bond strength and pack integrity. Both deviations from the target quantity — too little and too much — produce defects, but with very different signatures.

Under-application leaves an insufficient adhesive film between the joined panels. The bond appears intact at the folder-gluer, since the panels remain in contact as they exit the press section, but the actual contact area is reduced. Under transport stress, vibration or stacking load, the bond fails. The defect typically manifests downstream as pop-opens at the case erector, at the filling line or, in the worst case, at the end customer. Causes include partial nozzle blockage, drops in supply pressure during acceleration phases, and viscosity-induced flow restriction at low temperatures.

Over-application has the opposite signature. Adhesive squeezed out of the joint when the panels are pressed together migrates onto adjacent surfaces, where it causes strike-through, contamination of the print, or — on the inner surface — sticking of opposing panels in the flat-folded pack. On an absorbent recycled liner, over-application further extends the open time, since more water must migrate into the substrate before the bond reaches handling strength, which can lead to bond failure when the pack reaches the press section before the adhesive has set.

Nozzle build-up and standstill bonding

Water is the solvent of the PVAc dispersion. At the nozzle tip, it evaporates continuously, even during brief production interruptions. The polymer particles concentrate, coalesce and form an adhesive film at the nozzle opening. During extended standstills, the nozzle can become completely blocked. On restart, the dried film partially detaches and is ejected as particles onto the substrate, or the nozzle sputters irregularly until the channel is clear again.

The problem is compounded by frequent format changes: every stop creates a phase of nozzle drying. In operations with many short runs and frequent changeovers, nozzle build-up is one of the most common causes of adhesive application defects in the first blanks after restart. Tailing and satellite ejection accelerate the effect, as each microdroplet returning to the nozzle tip adds material to the drying film.

Glue-bead misplacement

The position of the glue bead relative to the crease line on the glue tab is a critical but frequently underestimated parameter. When the glue bead falls within the crease zone, or so close to it that the adhesive is pressed into the crease during joining, two problems arise. First, the crease — which serves as the hinge for folding — is stiffened by the adhesive. The pack cannot be cleanly erected at the case erector. Second, adhesive can be forced through the crease to the opposing side, creating a spot bond that fixes the panel in place.

With narrow glue tabs, as the PPWR increasingly demands [26], this problem becomes more acute. The margin between the crease line and the glue bead narrows, and even minor positional deviations result in overlap. Where a 35 mm tab still offered 10 to 15 mm of clearance, a 22 mm tab leaves only a few millimetres of tolerance.

Unintended panel bonding through adhesive misapplication

Splatter, satellites, tailing residues and squeezed-out excess adhesive can cause two panels of the flat-folded pack to bond at isolated points. These are known as inside gluers, where adhesive on the inner surface bonds opposing panels, or outside gluers, where adhesive on the outer surface bonds the blank to the one above or below it in the stack. At the case erector, these unintended bonds are critical: the machine attempts to erect the pack, but the spot bond resists.

In many cases, the opening force is sufficient to break the unintended bond, but at a cost. The separation causes surface damage: the liner delaminates, the print is destroyed, and the pack's visual appearance is ruined. For printed premium packaging, this is equivalent to waste. For unprinted transport packaging, the problem is frequently overlooked until the end customer complains.

9.2 Causes: Why Application Defects Occur

The defect patterns described in the preceding section can be traced back to a limited number of causes.

Causes of adhesive application defects and their associated patterns.

Low valve closing force	**Slow closing time → long adhesive thread → thread breakage with microdroplets**	**Satellites, tailing, splatter adjacent to the glue bead**
Viscosity variation	Too low → needle chatter; too high → extended thread formation on closing; shear-induced change during operation	Satellites, splatter, variable droplet size, irregular stitching pattern
Nozzle drying during standstill	Water evaporation → PVAc coalescence at nozzle tip → partial or complete blockage	Missing or irregular adhesive application after restart; particle ejection
Incorrect bead position	Nozzle misalignment relative to tab edge → adhesive in crease zone	Erection problems; strike-through into crease; stiffened fold hinge
Contamination in adhesive reservoir	Corrugated board debris, dust or foreign particles dissolve → altered flow behaviour	Sudden nozzle blockage; irregular adhesive flow; variable droplet size

One factor that is generally negligible in practice is adhesive supply pressure. Modern double-acting piston pumps inherently generate pressure pulsations, but a downstream pressure regulator reliably smooths them. A pressure characteristic stored in the control software automatically adjusts the material pressure to the machine speed. Pressure fluctuation as a cause of application defects is therefore no longer a real concern at the current state of the art.

By contrast, reservoir hygiene is far more underestimated. In practice, corrugated board debris, dust particles or film fragments are occasionally found in the adhesive supply vessel, where they dissolve in the aqueous dispersion. The result is altered flow behaviour and, more difficult to diagnose, intermittent, temporary blockages in individual nozzles that disappear after cleaning.

Viscosity: The Underestimated Variable

Adhesive viscosity influences satellite formation in two opposing ways. At low viscosity, the adhesive provides insufficient damping in the valve. The valve needle strikes the valve seat and rebounds. The adhesive surrounding the needle in the valve chamber offers too little resistance. This chattering generates uncontrolled adhesive release.

At high viscosity, by contrast, the adhesive retards the needle movement. As closing time increases, the adhesive thread becomes longer, and thread-breakage research shows that longer threads release more satellite droplets during break-up.

The recommended adhesive viscosity for clean application is 800 to 1500 mPa·s — a first approximation that holds for the majority of applications in practice. Some corrugated board operations use significantly lower viscosities, either for cost reasons or because other adhesive application systems in the plant have weaker valves that cannot process higher viscosities. A single viscosity standard across all systems is understandable from a business perspective, but it is an application-technology compromise that frequently comes at the expense of application quality at the folder-gluer.

A further aspect that qualifies the static viscosity measurement is the shear sensitivity of PVAc dispersions. Within the adhesive system, they undergo dynamic viscosity changes that depend on different factors. The valve needle moves faster as the machine speed increases, and the local shear rate in the valve increases accordingly. At the same time, the pressure regulator adjusts the material pressure to speed, altering the flow rate and therefore the shear loading throughout the pipework.

The nominal viscosity as measured with a Brookfield viscometer at constant rotational speed, therefore, describes the actual flow behaviour in the valve only incompletely. Where satellites or tailing occur with increased frequency despite nominally correct viscosity, it is advisable to consult the adhesive supplier. A targeted switch to a formulation with an adapted shear viscosity profile can eliminate the defect pattern. If the defect frequency changes after the switch, shear viscosity is the cause.

Static viscosity is also temperature-sensitive, and that sensitivity is what carries the variable across the production day. Water-based PVAc dispersions follow approximately Arrhenius rheology, with apparent activation energies in the range of 25 to 35 kJ/mol [31, 32]. In practical terms, viscosity rises by roughly 2 to 3 per cent per degree Celsius drop, and falls correspondingly when the hall warms up. A folder-gluer hall that is conditioned only by ambient air and the machinery's own thermal loading typically swings between roughly 18 °C in the early morning and 24 °C in the late afternoon. The viscosity of an adhesive nominally specified at 21 °C therefore drifts within a single shift across most of the working window, as Table 9.2 makes explicit.

Typical PVAc dispersion viscosity drift in an unconditioned folder-gluer hall (21 °C ± 3 °C). Nominal Brookfield viscosity 1100 mPa·s at 21 °C; actual values scale with the dispersion's specified viscosity.

18 °C — cold morning, unconditioned hall	× 1.20 to × 1.30	≈ 1320 to 1430	Long adhesive thread on closing; satellites and tailing rise; valve approaches its high-viscosity limit
21 °C — reference (typical spec point)	× 1.00 (reference)	≈ 1100 (nominal)	Centre of the working window; clean stitching, low satellite count
24 °C — warm afternoon, unconditioned hall	× 0.78 to × 0.85	≈ 860 to 935	Needle damping reduced; needle chatter, irregular droplet size, lower bond strength at low dose

Sources: viscosity-temperature behaviour of water-based PVAc dispersions per Pizzi & Mittal [31] and Schneberger [32]; activation energy assumed at 30 kJ/mol ± 5 kJ/mol, hence the range in column 2. The 800 to 1500 mPa·s working window cited above is exited at both ends within an ordinary day in any hall not actively conditioned.

Two operational consequences follow. First, the cold morning is a more dangerous regime than the warm afternoon: the viscosity rise increases the satellite count and the tailing length on every segment, and the worst applications of the day are typically the first two hundred blanks after start-up. Second, the shear-corrected viscosity at 24 °C is no longer in the centre of the window; it sits near or below the lower limit, and the valve loses the damping that suppresses needle chatter. The defect frequency is bimodal across the day rather than uniform, and the shift leader who reads only one viscosity measurement at a given time of day captures only half of the problem.

The critical point is that these causes interact. A valve with low closing force that also operates with an adhesive whose viscosity fluctuates with temperature throughout the day generates sporadic defects that are difficult to reproduce and, therefore, to diagnose. Only the combination of capable hardware, continuous monitoring, consistent reservoir hygiene and an adhesive viscosity matched to the valve system closes the control loop.

9.3 From the Machine to the End Customer: How Application Defects Propagate

The consequences of faulty adhesive application are not confined to the folder-gluer. They act as a cascade through the entire downstream value chain.

At the machine. Adhesive splatter and satellites contaminate not only the substrate but also machine components — conveyor belts, folding elements, and nip rollers. These deposits are transferred to adhesive-free surfaces on subsequent blanks, generating secondary surface damage unrelated to the actual bond. Cleaning intervals shorten, and on high-volume runs, the effect compounds.

At the case erector. Inside gluers and outside gluers can block the erection process. A single misglued pack can stop the erector and bring the entire upstream filling line to a standstill. At ten to twenty such incidents per shift, the accumulated downtime runs to hours, frequently classified as a box problem even though the cause lies in the adhesive application.

In the stack and on the pallet. By the time of palletising, the adhesive is generally fully set, so bonding between stacked packs rarely occurs. Surface contamination from splatter, however, remains a visible defect: adhesive spots on printed surfaces and liner damage from unintended, separated bonds are irreversible and are identified and rejected by the end customer at the latest.

At the end, the customer. Visible adhesive traces on the outer surface of the pack, liner damage from separated misglue bonds and dimensional deviations caused by a stiffened crease hinge all generate complaints. Chargebacks and supplier ratings react sensitively to defects of this kind, both optical and functional.

The deceptive quantity effect. Application defects are by nature minute quantities: a satellite weighs fractions of a milligram, a trailing thread barely more. This invites the assumption that the quantities are negligible. The opposite is true. A single adhesive spot of a few millimetres in diameter is sufficient to bond two panels that the case erector can no longer separate. The bond strength of a PVAc satellite far exceeds the liner's tensile strength: very little is required to cause significant damage.

The hidden risk. Many application defects are intermittent. They occur after restarts, at certain viscosity ranges and machine speeds, but not reproducibly. Without uninterrupted monitoring, they go undetected until the complaint arrives.

9.4 Remedies: High-Performance Application Technology

The most effective prevention against application defects begins at the valve. The performance of the adhesive application valve determines how cleanly a bead is deposited and how precisely a segment is terminated.

Closing force and switching time

A current-generation high-performance modular valve, such as the PX 1000 [66], achieves a closing force approximately 42 per cent higher than that of conventional valves. The consequence is direct: the adhesive strand is sheared cleanly, thread formation is minimised, and satellite generation is suppressed. A switching frequency of 750 Hz at 20 bar and 1000 mPa·s permits precise stitching application without tailing, even at high machine speeds. Comparable specifications are achieved by current-generation valves from other vendors; the engineering principle — high closing force, short closing time, minimised free volume around the needle — is what matters, not the badge on the housing.

A further advantage of high closing force is increased tolerance to viscosity variation. Where a low-performance valve fails at elevated viscosity, with the thread growing longer and satellites increasing, a high-performance valve continues to operate cleanly. This is decisive in practice and connects directly to Table 9.2: a valve specified at 1000 mPa·s but operated at 1400 mPa·s on a cold morning either tolerates the drift or it does not. A valve that functions only within a narrow viscosity window generates sporadic defects that are difficult to diagnose.

Water bath: protection during standstill

The most elegant solution against nozzle drying is an integrated water bath. The nozzle tips are automatically lowered into a water bath during production interruptions. The water prevents evaporation of the dispersion water at the nozzle tip, thereby preventing PVAc coalescence. On restart, the nozzles are immediately ready for use, without cleaning, misfires, or ejected particles.

For operations with frequent format changes and short runs, the water bath is a decisive factor in process stability. The alternative — manually cleaning the nozzles at every stop — is time-consuming, error-prone and unrealistic in practice during brief interruptions.

Purge firing: active nozzle cleaning

As an alternative or complement to the water bath, high-performance valves offer the option of purge firing: the valve switches rapidly, typically at 50 Hz, delivering a sequence of adhesive dots. The decisive effect here is not only the flushing of the nozzle channel; the rapid opening and closing generate a pumping action that dislodges dried adhesive from the channel and nozzle opening and expels it outward. The cleaning action, therefore, results not from the adhesive flow alone but from the mechanical-hydraulic pumping effect of the oscillating valve.

In most cases, it is sufficient to direct this cleaning sequence to the first blank after restart: the adhesive is applied where it is needed anyway, and the cleaning effect is a by-product. After an extended standstill, the sequence can alternatively be directed into a collection vessel or onto a waste substrate.

9.5 Adhesive Monitoring: From Presence Detection to Bead Position

Even the best application technology cannot eliminate defects entirely. It can, however, eliminate the risk of defects going undetected by enabling continuous monitoring of every applied bead.

Additive-free detection

Conventional adhesive monitoring is based on UV fluorescence: optical brighteners are added to the adhesive, fluoresce under UV light and are detected by a UV camera. This principle has three disadvantages.

First, the brightener additives are not free. Vendor data sheets cite an adhesive cost premium of up to 50 per cent for additised dispersions [67]. The figure depends strongly on the additive class. The cheap class, blue-fluorescing optical brighteners of the stilbene family, adds only modest cost but suffers a real detection problem: most coated white liners contain blue-fluorescing brighteners in the surface coating themselves, and the substrate fluoresces in the same band as the adhesive. The sensor's contrast collapses on exactly the substrate the operator most wants to monitor.

The expensive class, yellow- or green-fluorescent additives, or food-contact-grade chelates, give much cleaner contrast on white-top liner but justify the upper end of the 50 per cent premium. The choice within the additised path is therefore between cheap-brightener-on-kraftliner and expensive-brightener-on-coated-white-liner; the additive-free infrared route avoids the choice altogether.

Second, fluorescent additives are not permitted in some food-contact applications.

Third, not every dispersion is compatible with fluorescence additives, so the brightener requirement restricts the choice of adhesive.

Modern sensors operating in the near-infrared range, exemplified by the PGD 1000 and GDX 1000 [66, 67], detect adhesive based on its optical properties within a particular spectral band, without any additive. Detection is independent of the substrate background: printed or unprinted, light or dark, kraftliner or coated liner. The sensor distinguishes adhesive from substrate by their differing spectral signatures.

Bead-position monitoring

Presence detection alone answers only half the question: Is adhesive present? The second, equally important question is where, precisely, the bead has landed. Modern sensor systems simultaneously monitor up to three adhesive beads and determine their positions relative to the tab edge.

This becomes critical as tabs become narrower under the PPWR pressure documented in Section 9.1. On a 22 mm tab, the tolerance between the crease line and the tab edge is only a few millimetres. A 3 mm drift toward the crease line can lead to the erection issues described in Section 7.1. Without bead-position monitoring, this drift is missed: the pack appears correct but fails at the case erector.

Detection of contamination

A particular advantage of modern multi-band sensors is their ability to confirm reliably that adhesive-free areas are indeed free of adhesive. This sounds straightforward, but it is not: satellites and splatter are often only a few millimetres in size. Current sensors detect adhesive spots approximately 1 × 1 mm in size at sampling frequencies of 10 to 20 kHz [66, 67] — fine enough to capture the splatter that a case erector would catch a few minutes later, but on the wrong machine.

Comparison of adhesive detection methods.

	UV	IR
Additives required	Yes (optical brighteners). Cheap stilbene-class brighteners add modest cost but contrast poorly on coated white liners; food-contact-grade or non-stilbene brighteners drive cost up to roughly +50 % vs. additive-free dispersion (vendor data sheet, Baumer hhs [67]).	No; any dispersion can be used.
Substrate dependency	High. Coated white liners frequently contain blue-fluorescing brighteners in the surface coating, so the substrate fluoresces in the same band as a stilbene-additised dispersion and the sensor's contrast collapses.	Low. Functions on all common substrates: kraftliner, white-top, coated, printed, recycled.
Food-contact suitability	Restricted; some brighteners are not permitted in direct food-contact applications.	Unrestricted; no additive in the adhesive.
Bead-position measurement	Yes, depending on system.	Yes; up to three beads simultaneously, with position relative to tab edge.

Splatter detection	Only with sufficient additive concentration and substrate contrast.	Spots from ≈ 1 × 1 mm at sampling rates of 10 to 20 kHz [66, 67].

9.6 A Three-Level Safety Concept

The individual measures — high-performance valve, water bath, purge firing, multi-band sensor monitoring, SPC analysis — only achieve their full effectiveness when combined, as each step supports the others. The three-level safety concept consists of interconnected layers that collectively raise application reliability.

Level 1 — Prevention. A high-closing-force valve at high switching frequency (the PX 1000 at up to 750 Hz [66] is one worked example) ensures clean, tailing-free application. The water bath protects the nozzles during standstill. Purge firing cleans the channel on restart. Together, these measures eliminate the most common causes of application defects.

Level 2 — Detection. A multi-band IR sensor (PGD 1000 or GDX 1000 [66, 67] are the worked examples) monitors every applied bead in real time: is adhesive present, is the bead in the correct position, and are adhesive-free areas genuinely free? Defects are identified before the pack leaves the folder-gluer.

Level 3 — Response. Detected defects trigger an immediate reaction — defective blanks are marked or ejected. Where consistent deviations occur, such as bead-position drift, the control system can readjust the nozzle position. Beyond the per-blank reaction, the data stream feeds into the SPC tools described in Section 9.7, which raise alarms on trends rather than thresholds and enable predictive maintenance. The Response level in this architecture is both reactive and predictive.

Integrated platforms such as CorrBox Solution [65] combine application and monitoring within a shared control environment, and an SPC layer is the natural extension of that architecture. The reported adhesive savings of up to 60 per cent through optimised stitching are a welcome side-effect; the primary value lies in the closed control loop the integrated platform enables.

9.7 Summary

Adhesive application quality is not a secondary topic. It is an independent dimension of packaging performance, separate from washboarding, warp, and crease quality, and capable of stopping the case erector even when all upstream parameters are within their windows. Six defect patterns recur: satellites and splatter, tailing, under- and over-application, nozzle build-up during standstill, glue-bead misplacement, and unintended panel bonding through misapplication, each with a distinct signature and downstream consequences.

The causes are understood physically: viscoelastic thread formation at the closing valve, dispersion drying at the nozzle, positional tolerance against an ever-narrower tab, and shear- and temperature-driven viscosity drift across the production day (Table 9.2). High-performance valves with high closing force and short closing time address the primary causes; the valve specification must be matched to the viscosity range that the hall climate will actually deliver, not the one printed on the data sheet.

Prevention alone is not sufficient. Modern multi-band IR sensors enable 100 per cent monitoring, presence, position, splatter, without the need for adhesive additives, even on substrates such as coated white liner where stilbene-based brightener systems lose contrast. For narrow tabs under the PPWR, this monitoring is no longer optional.

The three-level safety concept, prevention (valve, water bath, purge firing), detection (multi-band IR sensor), response (bundle or per-blank ejection), reflects the current state of the art. When the IR sensor coexists with inline waterscoring, the wavelength interaction documented in Section 9.6 must be considered during integration.

The preceding chapters have shown how washboarding, warp, liner type and crease quality influence the bond at the manufacturer's joint. Yet even when all these parameters are within their optimal ranges, a single defect class can still bring the entire process to a standstill: faulty adhesive application. The signatures are familiar from the shop floor: splatter on supposedly adhesive-free surfaces, satellite droplets, misplacement of the glue bead relative to the crease line, and tailing threads between segments.

These defects originate in the application technology, but they propagate through material physics — viscoelastic thread formation, dispersion drying, and surface tension at the nozzle. They strike the value chain at a particularly vulnerable point: the case erector. This chapter analyses the defect patterns, their causes, their consequences and the technical solutions available today to prevent and monitor them.

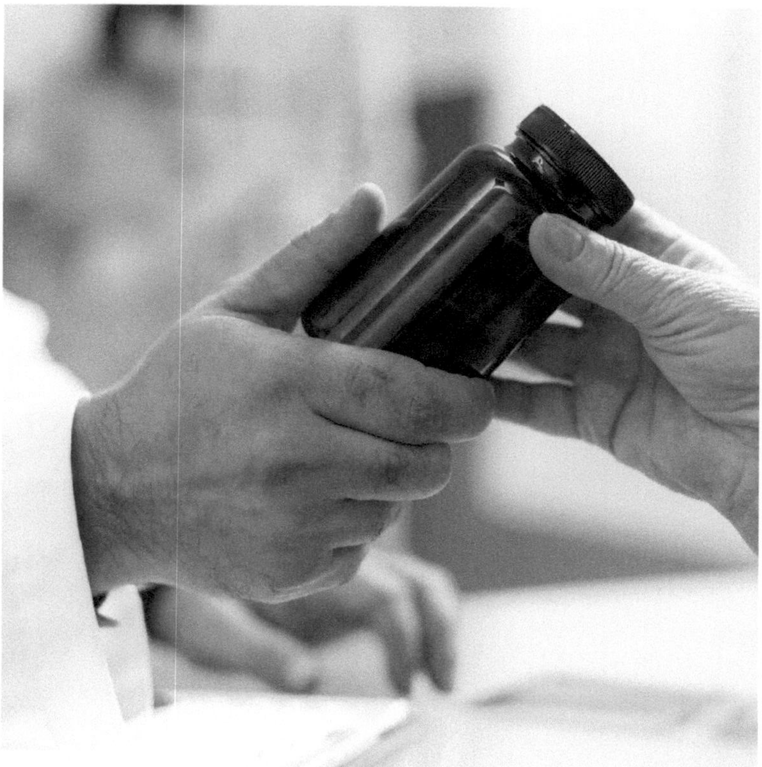

Part IV - The Treatment (Therapy)

Therapy follows diagnosis. Part IV presents two interventions and their extension beyond the folder-gluer. Neither is a miracle cure. Each addresses a specific failure mechanism documented in Parts I-III and is grounded in published evidence.

Chapter 10: Stitching: Damage Tolerance Over Maximum Strength

Introduction card

A stronger joint does not always require more adhesive. This chapter explains why interrupted, offset adhesive segments can reduce consumption while making the manufacturer's joint more damage-tolerant. The effect depends on redundancy: stitching stabilises a three-bead network, but can destabilise a two-bead one. It is relevant for process engineering, adhesive specification, procurement, and capex planning.

"Can you make a glue line stronger by removing material in a structured pattern, and why does the same intervention that destabilises a two-bead joint stabilise a three-bead one?"

Economic impact: Stitching can reduce adhesive consumption at the manufacturer's joint by up to 50 per cent at design level. Test data on B-flute show that a three-bead continuous reference reached 9,350 N/m, while three beads with 30 per cent stitching reached 16,396 N/m. At 50 per cent stitching, strength remained higher than the reference at 13,562 N/m. Removing one bead entirely produced the opposite result, with a strength loss instead of a gain.

Operational efficiency: Stitching can be implemented as a software change on current-generation valve and controller systems. Legacy folder-gluers may require a head swap or valve upgrade. At 400 m/min, the valve-switching frequency is about 270 Hz, within the capability of modern high-closing-force valves. Successful implementation depends on three details: real interruption of the bead, lateral offset between adjacent tracks, and adhesive at the leading and trailing ends of each track.

Quality: Stitching changes the failure mode. A continuous bead can allow one crack to propagate along the joint. A stitched bead forces each segment to fail separately, while the gap acts as a crack arrester. This changes the strength distribution from one governed by the weakest point to one governed more by the mean. In the cited data, the coefficient of variation fell from 0.37 to 0.07 at 30 per cent stitching.

Ecological impact: Lower adhesive use at equal or higher reliability reduces the embedded carbon and water linked to PVAc production. On multi-million-pack runs, the saving becomes material.

Limitations: The available data are on B-flute. The behaviour on heavier flutes has not been published. Stitching also requires redundancy. Removing one bead and stitching the remaining two can destabilise the joint rather than improve it.

Read in full: process engineers, adhesive specifiers, plant managers, anyone planning capex.

Card sufficient: procurement teams and executives reviewing the consumables cost case.

Prerequisite: Chapter 9.

Follow-up: Chapter 11, waterscoring as the load-reduction layer that multiplies the effect of stitching.

Chapter 6 showed why conventional parameter optimisation reaches its limits: variation is uncontrollable, interactions are multidimensional, and external drivers narrow the available room for manoeuvre. The logical consequence is a shift in thinking, away from maximising individual strength values and towards the robustness of the overall system. This chapter introduces the first of the two interventions: stitching, the transition from continuous adhesive beads to offset, interrupted application patterns at the manufacturer's joint. The principle does not originate in the packaging industry but in a field where failure costs lives — aviation.

10.1 The Principle: Crack Arrest Rather Than Maximum Area

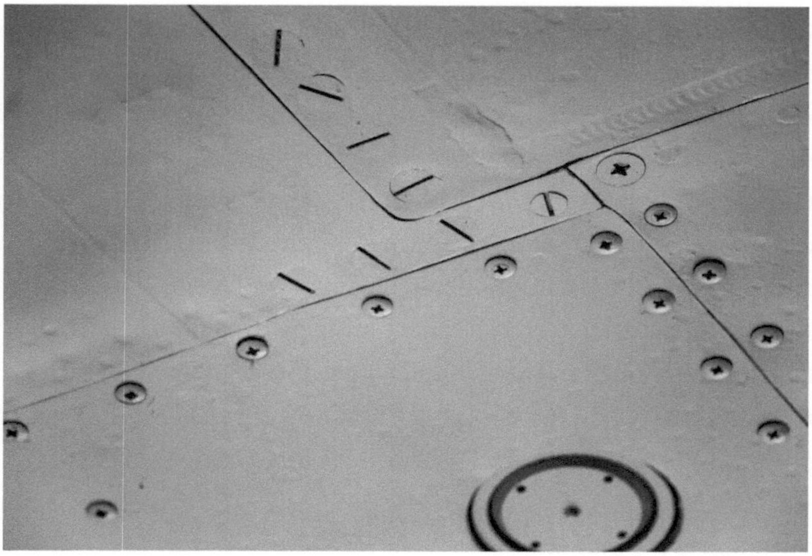

Figure 32 Aircraft skin showing rivet rows along panel seams and access-panel fasteners. Each rivet is a crack-arrest point: a fatigue crack initiated at one fastener has to reinitiate across unbonded material before it can reach the next. The structure is not stronger than a continuous weld would be, but it fails segment by segment rather than all at once. Stitching at the manufacturer's joint applies the same principle to corrugated board, at a different scale and in a different bonded material.

A lesson from aviation

On 28 April 1988, an Aloha Airlines Boeing 737-200 lost an entire fuselage section at cruising altitude. The aircraft landed with a roof panel missing. The cause: thousands of small fatigue cracks at rivet holes had linked into a continuous crack — a phenomenon known as multi-site damage. A single crack path with nothing to interrupt it.

The aviation industry already knew the problem. In 1954, two de Havilland Comets, the first jet airliners, broke apart because fatigue cracks at the corners of square window cut-outs propagated unimpeded through the fuselage.

The consequence was the development of fracture mechanics as an engineering discipline. Griffith [76] formulated the energy-based crack-propagation criterion; Irwin [77] translated it into the practically applicable stress intensity factor K; Paris and Erdogan [78] described the fatigue-crack growth rate as a function of the cyclic stress-intensity range ΔK. From these foundations, aviation developed damage-tolerant design (DTD): a structure is not designed on the assumption that no cracks will form — because that is unavoidable — but so that cracks cannot propagate uncontrollably. Rivet lines, stringers, frames and doubler plates act as crack arresters: they interrupt the crack path, force the crack to reinitiate, and give the inspection programme time to find the crack before it becomes critical.

From aviation to the glue line

What does an aircraft fuselage joint have in common with the manufacturer's joint of a corrugated box? More than the difference in scale suggests. Both are bonded joints under variable loading. Both are susceptible to fatigue and to localised defects. In both cases, the question is whether a local defect leads to total failure. The lesson from aviation translates directly.

A continuous adhesive bead at the MJ is the equivalent of an uninterrupted, seamless joint: it offers maximum strength under ideal conditions. But as soon as a defect occurs — a dry spot, an air pocket, an area of insufficient contact pressure — the crack can propagate along the entire bead without encountering an obstacle. The failure mode is propagation-driven: the question is not whether the crack initiates, but whether it can spread.

Stitching — the transition to offset, interrupted adhesive segments — transfers the crack-arrest principle to the glue line. Each gap between two segments acts as a geometric discontinuity: the crack would need to reinitiate across unbonded material, which requires additional energy. The failure mode becomes initiation-driven. It is no longer crack propagation but reinitiation at the next bonded segment that determines the failure load.

10.2 Why Less Is More: Failure Behaviour Under Realistic Loading

The difference between a continuous bead and stitching only becomes apparent under realistic loading. Under uniform compression in the laboratory — the BCT — the continuous bead is theoretically superior: a larger bonding area means greater shear strength on paper. But laboratory conditions do not reflect reality on the pallet, and the central argument of this chapter is that the loading mode in the laboratory is not the one that actually causes boxes to fail in the field.

Five loading scenarios

Scenario 1: Uniform compression (BCT laboratory). Under ideal conditions, the continuous bead forms a single buckling band extending across the full length of the MJ. With stitching, the deformation distributes across several smaller buckling zones between the bonded segments. The peak load of the continuous bead is higher, but the difference typically lies within five to ten per cent — well within the system's scatter.

Scenario 2: Point load on the pallet. Deck-board gaps, nail heads and broken wood fibres generate localised stress peaks. With a continuous bead, a local defect triggered by the point load can propagate along the entire bead because the crack path remains uninterrupted. With stitching, propagation is limited to the affected segment. The gap to the next segment serves as a crack arrester.

Scenario 3: Eccentric load from a misaligned MJ. A continuous bead fails at the most heavily loaded point and propagates along the bead in a cascading failure. Stitching fails at the most heavily loaded segment first, leaving the remaining segments to redistribute the load sequentially.

Scenario 4: Sustained load (creep). Boxes carrying a load for days or weeks experience slow crack growth. With a continuous bead, cracks grow uninterrupted; stitching forces cracks to restart at each gap, increasing service life under stack pressure.

Scenario 5: Dynamic and cyclic load. Vibration during transport produces cyclic fatigue. With a continuous bead, longer cracks grow faster — a self-reinforcing feedback loop. Stitching resets the crack at each segment and breaks the feedback loop.

The comparison is summarised in Table 10.1. One row in the table is highlighted in colour, and that highlight is deliberate. Of the eight criteria listed, the BCT-laboratory row is the only one in which the continuous bead wins, and it is also the one that most procurement specifications still treat as the primary acceptance criterion. The five-to-ten per cent BCT advantage of the continuous bead is the central objection of every BCT-believing buyer, and the rest of this chapter exists to address that objection on its own terms.

Failure behaviour — continuous bead versus stitching, by loading mode.

Bonding area	**100 % of bead length**	**50 to 70 %**
Stiffness profile	Continuously stiff composite zone	Alternating stiff and flexible
Crack path	Uninterrupted: propagation without obstacle	Interrupted: arrest at each gap
Failure mode	Propagation-driven	Initiation-driven
BCT (laboratory) — the buyer's objection	**0 to 10 % higher — the only column where the continuous bead wins**	**Reference — but the laboratory is not the pallet (Section 10.3)**
Point load (pallet)	Local defect propagates along the bead	Defect remains locally contained
Sustained load (creep)	Uninterrupted crack growth	Stepwise arrest at every segment gap
Dynamic / cyclic load	Self-reinforcing feedback (longer crack → faster growth)	Feedback loop broken at every gap

The table makes the central argument explicit: what looks like a disadvantage in the laboratory, a smaller bonding area, a slightly lower BCT, becomes an advantage under practical conditions, because it changes the failure mechanism rather than scaling it. The next section follows the highlighted row out of the laboratory and onto the pallet, where every other row of the table starts to apply at the same time.

10.3 The Weakest-Link Argument: Why Stitching Outperforms on the Pallet

The conditions on the pallet bear little resemblance to those in the BCT laboratory. In the laboratory, the box stands on a flat steel platen. On the pallet, it stands on deck boards with gaps, overhangs and surface irregularities. Kim et al. [43] show that an overhang of just 50 mm reduces BCT by 29-40 per cent. Modzelewska [10] documents a 20-40 per cent reduction in BCT at ambient humidity above 60 per cent RH.

Cumulatively, and this accumulation is the reality, the reduction factors reach 50 to 70 per cent of the nominal BCT. A box carrying 8 kN in the laboratory carries only 2.4 to 4.0 kN on the real pallet. Under these conditions, the last 5 to 10 per cent of bonding area does not matter; the question is whether a local defect leads to system failure.

The variation argument

Chapter 6 identified variation as the true enemy. The flute-phase position varies uncontrollably along the crease line. The washboarding amplitude fluctuates with starch add-on at the corrugator. Liner porosity varies from batch to batch. All of these sources of variation superimpose at the MJ, producing an unpredictable pattern of strong and weak zones along the glue line.

In a continuous bead, the weakest section determines the joint's strength according to the weakest-link principle. A single dry patch, a single air pocket, a single zone of insufficient contact pressure can start the crack that severs the entire bead. With stitching, each segment can fail independently. A weak segment does not impair the load-bearing capacity of the remaining segments. The overall load capacity is determined not by the minimum but by the mean segment strength, which is statistically far more stable than the minimum.

In short, stitching transforms the failure load from an extreme-value distribution into a normal distribution. The mean of a sample is well behaved; the minimum is not. The full statistical derivation, with the worked-out expectation $\mu - 1.54\sigma$ for the minimum of ten independent normally-distributed segments, is given in Footnote [1].

The rivet-network analogy

The aircraft manufacturer uses rivets, not because rivets are stronger than welds — they are not — but because a network of a hundred rivets contains a hundred crack arresters. When one rivet fails, the neighbouring rivets take up the load. The crack is stopped before it reaches a critical length. The structure is not stronger; it is more reliable.

This is precisely what stitching achieves at the MJ. It replaces a single continuous bond with a network of discrete bonded segments. Individual strength decreases marginally; system reliability increases substantially.

The same principle applies to dotted hot-melt patterns. The aviation analogy is indifferent to whether the discrete bonded zones are stitches of cold-glue dispersion or dots of hot-melt; what matters is that the bonded area is segmented and that the crack must reinitiate across unbonded material between zones. Chapter 13 develops this in detail for the hot-melt case (dot pitch, dot diameter, line speed, and the interaction with thermal mass at the closing nip). The reader

---

[1] Statistical derivation of the weakest-link argument. Consider a simplified model of a glue line comprising n independent segments with mean strength μ and standard deviation σ. The overall strength of a continuous bead is determined by its weakest segment; with $n = 10$ and a normal distribution, the expected value of this minimum is approximately $\mu - 1.54\sigma$ (the standard order-statistic result for the minimum of ten i.i.d. normal variates). The overall strength of a stitched pattern, by contrast, is determined by the mean of the segments minus the loss from the unbonded gaps. At a coefficient of variation $\sigma/\mu > 0.2$ — a realistic value at the MJ — the statistical advantage of comparing means rather than minima outweighs the area loss, and the stitched joint is stronger than the continuous one in expectation, not despite the smaller bonded area but because of the segmentation that goes with it.

who finds the logic of this chapter persuasive should treat Chapter 13 as the same argument applied in a different temperature regime, not as a separate one.

10.4 How the Stitching Pattern Is Generated in Practice

The argument so far has been mechanical: stitching arrests cracks, redistributes load, and converts an extreme-value problem into a mean-value problem. The remaining question is how the pattern is actually printed onto the tab at folder-gluer speed. Three implementation details determine whether the engineering principle survives contact with the production line.

Interruption alone is not enough — the offset between adjacent tracks.

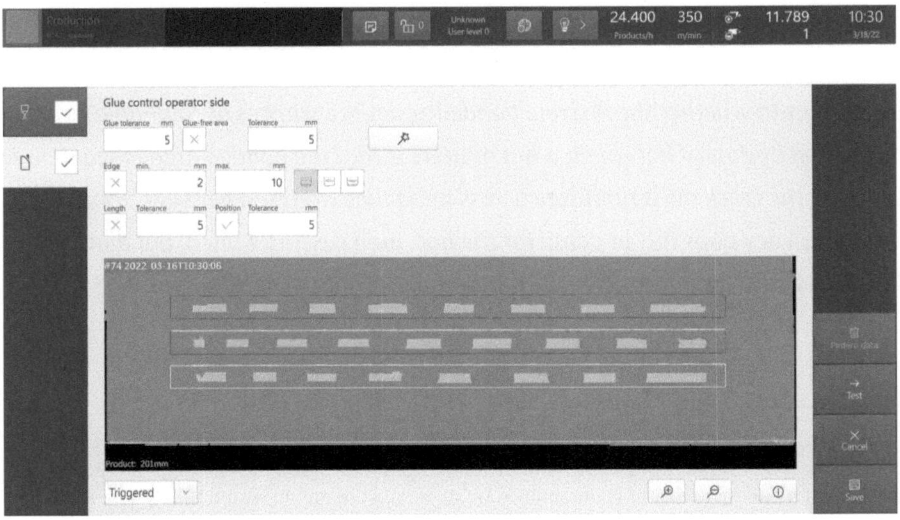

Figure 33 Production-line sensor view of a three-track stitching pattern at the manufacturer's joint. Two features carry the argument of this section. First, the beads in each track are interrupted into discrete segments — the stitching pattern of Section 10.1. Second, the gaps in adjacent tracks are offset against each other rather than synchronous: wherever one track has a

gap, the neighbouring track has adhesive, and the bonded fraction read across the tab width stays roughly constant at every longitudinal position. This is the offset condition without which the crack-arrest argument collapses into a row of synchronised weak bands.

Stitching, in its strict sense, is not just the interruption of a single bead. It is also the lateral offset between adjacent tracks. If the gaps in track 1 line up with the gaps in track 2 — if every adhesive-free zone runs straight across the full tab width — the crack arrest argument collapses, because the structure now contains a single transverse weak band at every gap position. The crack does not have to reinitiate; it has been handed a continuous unbonded line to follow.

The mechanically correct stitching pattern offsets the gaps in adjacent tracks, so that wherever one track has a gap, the neighbouring track has adhesive. Read across the tab width, the bonded fraction is roughly constant at every longitudinal position; read along any single track, it is interrupted. The crack that wants to propagate must do so against an alternating boundary condition: arrested in one track, supported in the next. This is the geometry that delivers the crack-arrest behaviour described in Section 10.1, not a row of synchronised gaps.

Every track must end with adhesive, not with a gap.

A second software detail matters at least as much. The leading and trailing ends of a track are exactly the regions a peel load attacks first — they are the free edges of the joint. If the software lets a track end on a gap, the last centimetre of the joint is unbonded at the very point where the peel stress is highest. The control logic in a properly engineered stitching system, therefore, guarantees that every track terminates with an adhesive segment, regardless of where the track-end falls in the cycle. The stitching pattern is end-anchored at both ends and interrupted only in the middle.

What hardware is needed

Within Baumer's portfolio, all variants of the CorrBox Solution platform [65] support stitching out of the box: track interruption, inter-track offset, and end-of-track adhesive guarantee are part of the standard control logic, with no upcharge or special configuration. Comparable platforms exist from other vendors. The relevant capability is not the brand but the combination of (a) a high-closing-force valve that can switch cleanly at the required frequency, (b) an integrated controller that can coordinate multiple tracks, and (c) the software that enforces the offset and the end-of-track rule.

The valve switching frequency is the binding constraint at high line speeds. At 400 m/min, the substrate moves at roughly 6.7 m/s. A 12.5 mm segment with a 12.5 mm gap (50 per cent stitching saving) repeats every 25 mm, requiring approximately 270 valve cycles per second. A current-generation high-performance valve, such as the PX 1000 — specified up to 750 Hz at 1000 mPa·s [66] — sits comfortably above this, with headroom for finer patterns or higher line speeds. A legacy valve with a switching limit of 100 Hz cannot deliver the pattern cleanly at 400 m/min; the segment ends smear, the offset is not maintained, and the engineering principle is lost on the production line, even though it is intact on the engineering drawing.

The implication for an investment decision is straightforward. A modern folder-gluer with a current-generation valve and an integrated controller delivers stitching as a software setting. An older folder-gluer requires either a head swap (legacy wheel applicator) or a valve upgrade (low-frequency valve) before the engineering principle can be realised at production speed. The cost of the upgrade is comparable to a few weeks of the adhesive savings it enables, but the timing of those savings differs between vendors and generations, not the principle itself.

10.5 Less Adhesive, Higher Strength: The Quantitative Evidence

Bonding area and strength

The reduction in bonded area through stitching is typically 30-50%. With a tab width of 30 mm and a bead width of 3 to 4 mm, a single continuous bead produces a bonding area of approximately 3 to 4 mm multiplied by the box length. With stitching at 50 per cent area coverage, this value is halved.

Under shear stress — the most favourable loading mode — this would imply a proportional loss of strength. But in practice, the MJ glue line is not primarily loaded in shear; it is loaded in peel and cleavage (Chapter 3). Under peel stress, the load distribution is concentrated along a narrow strip at the bead edge in any case, with the inner areas of the bead contributing little to bond strength. The effective strength reduction under peel is therefore considerably smaller than the area reduction, and, as the next subsection shows, the statistical effect of segmentation reverses the sign altogether.

Statistical effect and laboratory evidence

PTS test report 35.756 measured the Factory Edge Strength of B-flute specimens bonded in six different ways: three beads continuous (the reference), three beads with 30 per cent and 50 per cent stitching savings, two beads continuous (one bead removed), and two beads with 30 per cent and 50 per cent stitching savings. Ten specimens per condition. The full methodology, per-condition statistics and a boxplot of all six conditions are reproduced in Appendix 10. The headline numbers are summarised here.

The mean Factory Edge Strength rises from 9,350 N/m for the three-bead continuous reference to 16,396 N/m for three beads with 30 per cent stitching saving — an increase of 75 per cent at a 30 per cent reduction in adhesive consumption. At 50 per cent stitching saving, three beads, the mean is 13,562 N/m, still 45 per cent above the continuous reference. Removing one bead entirely — the alternative saving strategy — produces the opposite sign: the mean drops from 9,350 N/m to 8,730 N/m, a 7% loss, for what is nominally also an adhesive saving. Same savings, opposite direction.

The variance behaves the way the weakes tlink argument predicts, but only when stitching has a redundant network of beads to act on. The three-bead continuous reference has a coefficient of variation of 0.37; the three-bead 30-per-cent-stitching condition has a CV of 0.07. The scatter falls by a factor of five, while the mean rises by 75 per cent. The 50-per-cent-stitching condition lies between the two, at CV 0.14. All three of these conditions retain three independent bead networks; what stitching changes is the segmentation of each network, not their count, and the variance reduction is the empirical signature of the weakest-link-to-mean transformation argued in Section 10.3.

When the saving strategy is to remove a bead, the variance argument reverses. Two-bead continuous has CV 0.39, essentially indistinguishable from the three-bead continuous reference, because the dominant scatter in both cases is material-driven. Combining bead removal with stitching makes the variance worse, not better: two beads with 30-per-cent stitching have a CV of 0.42, the highest of all six conditions in the test, and two beads with 50-per-cent stitching have a CV of 0.30, with a mean only 28 per cent above the reference. The variance argument, therefore, protects three beads with stitching, not stitching as a free-standing intervention. Removing a bead removes the redundancy on which crack arrest depends; the segmentation that stabilises a redundant network destabilises a thinned one.

One observation from the test report is at least as important as any single number in the table. In all, 60 individual specimens across all 6 conditions, the bond at the manufacturer's joint did not separate. Failure was a material failure throughout: tears in the flutings and liners, not delamination at the glue seam. The PTS report's own conclusion is that the bond is stronger than the substrate it joins, on this material. The high variance in the continuous-bead conditions, therefore, reflects substrate variation, not bond variation, and the variance reduction in the three-bead-plus-stitching conditions is the signature of a segmentation pattern that lets the load find more paths through that variable substrate before the substrate breaks. Stitching does not so much strengthen the bond as make the substrate behave as if it were less variable.

One qualifier on the "bond stronger than substrate" conclusion belongs here, however. The PTS test was conducted on a B-flute board, which is the most common flute in commercial practice and a fair representative of the regime in which most secondary packaging is produced. On heavier flutes, C-flute, BC- and BA-double-walls, EB- and AB-combinations, the substrate offers more cross-section to resist the FES test and is correspondingly stronger; whether the bond remains stronger than the substrate at those flute weights, or whether the manufacturer's joint emerges as the limiting weak point, is not answered by this data set. The reasonable expectation is that on the heaviest flutes, the MJ will become measurable in the test — the substrate's strength grows faster with flute height than the glue seam's does — but the experiment to confirm that has not been published. The B-flute conclusion should not be extrapolated to the heaviest grades without test confirmation.

The PTS investigation, therefore, demonstrates that stitching does not merely save adhesive; it increases the assembly's strength while improving load distribution. Three beads at 30 per cent stitching saving deliver +75 per cent in mean Factory Edge Strength and a five-fold reduction in coefficient of variation, simultaneously, against three beads continuous. That joint result is the experimental answer to the BCT-laboratory objection of Section 10.2. The full data on which this answer rests is the subject of Appendix 10.A;

the boxplot in Figure 10.A.1 makes the joint result visible at a glance, and the table beneath it records the per-condition numbers.

10.6 Summary

Stitching is not a way to make a stronger glue line by adding material. It is a way to make a more reliable glue line by removing material in a structured pattern. The principle is borrowed from aviation, where the difference between a continuous joint and a segmented one was paid for in human lives before it was paid for in test machines.

The mechanical argument runs through four steps. First, a continuous bead fails by crack propagation; a stitched bead fails by reinitiation. Second, on the pallet, the loads that matter are point loads, eccentric loads, sustained loads and cyclic loads — every column of Table 10.1 except BCT. Third, the BCT advantage of the continuous bead, the only column it wins, is five to ten per cent in a domain where overhang and humidity together strip 30 to 50 per cent. Fourth, stitching converts the strength distribution from an extreme-value distribution to a normal distribution, so the mean rather than the minimum determines field performance.

The implementation argument runs through three details. The interruption must be combined with a lateral offset between adjacent tracks; every track must end with adhesive rather than a gap; and the valve switching frequency must comfortably exceed the cycle rate required by the line speed. All three details are handled in software on a modern folder-gluer with an integrated controller; on older generations, they require either a head swap or a valve upgrade. The CorrBox Solution platform supports stitching across all variants; comparable platforms from other vendors do the same.

The empirical argument is in the PTS test report 35.756: against the three-bead continuous reference, three beads with 30 per cent stitching saving deliver +75 per cent in mean Factory Edge Strength and a five-fold reduction in coefficient of variation, simultaneously. Across all sixty B-flute specimens, the bond at the manufacturer's joint never separated; substrate failure was observed throughout. That last finding bounds the conclusion: on B-flute, the bond is the stronger link of the assembly, and the variance reduction recorded here is the signature of a segmentation pattern that lets the load find more paths through a variable substrate. On heavier flutes, the relative positions of the bond and the substrate may invert, and that experiment has not been published. Within the regime in which the test was actually run, the data is the experimental answer to the BCT-laboratory objection of Section 10.2; outside that regime, the answer is a reasonable expectation rather than a measurement.

Appendix 10.A: Test Methodology and Data — PTS Report 35.756

This appendix documents the test on which the +75 per cent Factory Edge Strength finding of Section 10.5 rests. The test was conducted at PTS (Papiertechnische Stiftung) and is recorded as test report 35.756. The methodology is reproduced here in the detail that the body of the chapter cannot carry, so that the finding can be evaluated against its own evidence rather than taken on trust.

A.1 Test method

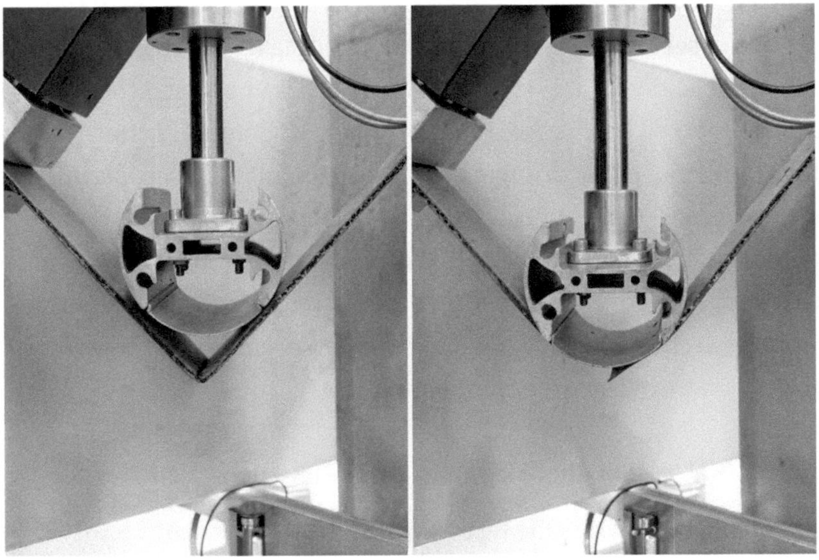

Figure 34 Factory Edge Strength test apparatus per FEFCO 55 / DIN 19304. The specimen is pre-folded along the manufacturer's joint into a V-shape and clamped at the apex; a steel test cylinder is then driven through the V opening at a constant 12.5 mm/min, applying a peel load that opens the V apart at the bonded apex. Left: just before the cylinder engages the inner face of the V. Right: after separation at the manufacturer's joint. The reading is the force at separation divided by specimen width, expressed in N/m. (Quelle PTS)

Factory Edge Strength was measured according to FEFCO test method No. 55. Rectangular specimens 100 mm wide and 400 mm long were cut from production boxes at equal distances from the manufacturer's joint. The specimens were clamped perpendicular to the MJ and loaded at a constant feed rate of 12.5 mm/min until the glue seam failed. Failure load was recorded and divided by specimen width to give the strength in N/m. Ten specimens were tested per condition (n = 10).

A.2 Conditioning

All specimens were conditioned for a minimum of 24 hours under standard conditions in accordance with DIN EN ISO 187:2003-02 at 23 ± 1 °C and 50 ± 2 per cent relative humidity, and tested in the same climate. The conditioning is essential because linear tensile strength and fracture toughness depend strongly on moisture (see Section 8.5 and the moisture-sensitivity values reported by Cornaggia et al. [3] and Szewczyk and Głowacki [15]); without controlled conditioning, the differences between conditions could not be attributed to the bond geometry.

A.3 Production conditions

All specimens were B-flute boxes produced under real production conditions at a folder-gluer line speed of 200 m/min. The line, the substrate batch, and the adhesive batch were held constant across the four conditions, so that the only variables were the application pattern — number of beads (two or three) and the stitching ratio (continuous, 30 per cent saving, or 50 per cent saving). Holding the production environment constant is what makes the comparison between conditions possible at all.

A.4 Per-condition statistics

Factory Edge Strength by condition (PTS 35.756, FEFCO 55, B-flute, n = 10 per condition).

Sample 1 — 3 beads, continuous (reference)	10	9 350	3 469	0.37	4 008 / 7 008 / 9 350 / 11 692 / 14 692

Sample 2 — 3 beads + 30 % stitching	10	**16 396**	1 229	0.07	14 503 / 15 566 / 16 396 / 17 226 / 18 289
Sample 3 — 3 beads + 50 % stitching	10	**13 562**	1 912	0.14	10 618 / 12 271 / 13 562 / 14 853 / 16 506
Sample 4 — 2 beads, continuous	10	8 730	3 438	0.39	3 435 / 6 409 / 8 730 / 11 051 / 14 025
Sample 5 — 2 beads + 30 % stitching	10	10 472	4 353	0.42	3 768 / 7 534 / 10 472 / 13 410 / 17 176
Sample 6 — 2 beads + 50 % stitching	10	11 965	3 576	0.30	6 458 / 9 551 / 11 965 / 14 379 / 17 472

Means and standard deviations as reported in the PTS test report 35.756. Five-number summaries (min / Q1 / median / Q3 / max) computed from mean and SD using a normal-distribution approximation; the per-specimen raw values are held by PTS and were not published in the report. The CV column is the dimensionless ratio SD / mean and gives the cleanest at-a-glance comparison of scatter across conditions. Note that in all sixty individual tests, the bond at the manufacturer's joint did not separate — substrate failure (tear of flutings and liners) was observed throughout — so the variance recorded here is dominated by substrate variation rather than bond variation. See Section A.7 on the implications of that finding for heavier flutes.

A.5 Boxplot

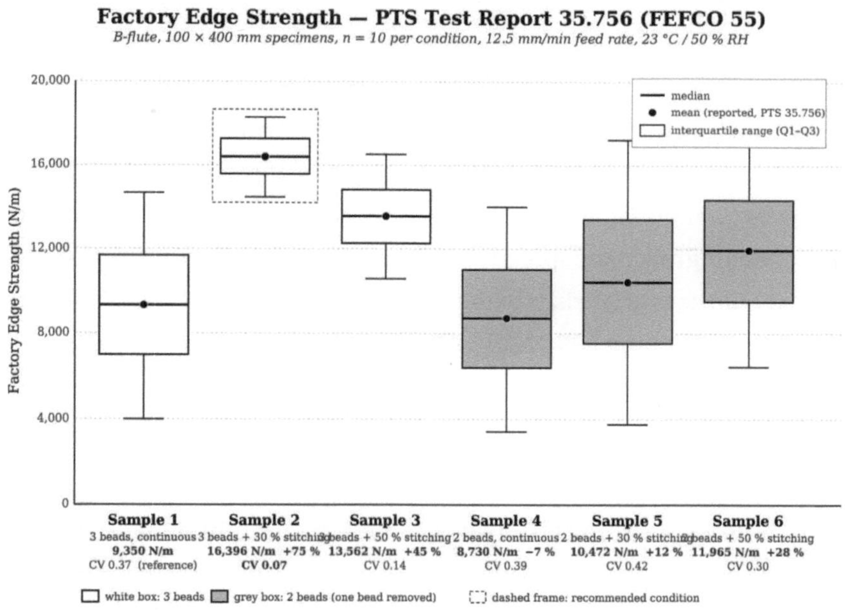

Figure 35 Factory Edge Strength across all six PTS 35.756 conditions (B-flute, n = 10 each). White boxes mark the three-bead conditions; grey boxes mark the two-bead conditions in which one bead has been removed. Box: interquartile range (Q1–Q3). Horizontal line: median. Solid dot: reported mean. Whiskers: extended range (computed min–max from mean and SD). Dashed frame: recommended condition (3 beads + 30 % stitching saving).

A.6 Reading the boxplot

Four observations follow directly from the boxplot. First, the recommended condition — Sample 2, three beads with 30 per cent stitching saving — sits in a class of its own. Its interquartile range does not overlap with any other condition, its mean (16,396 N/m) is the highest in the test, and its scatter (CV 0.07) is the tightest in the test. The dashed frame in the figure marks this condition not for visual flourish but because the joint result — highest mean, lowest variance — is what the engineering argument predicts and what every other condition fails to reproduce.

Second, the comparison between the three-bead and two-bead families tells the chapter's central story in two opposite directions. Within the three-bead family (Samples 1, 2, 3), stitching delivers exactly what Section 10.3 predicts: the boxes shift up and tighten as the segmentation progresses from continuous to 30 per cent saving, then to 50 per cent saving. Within the two-bead family (Samples 4, 5, 6), stitching does not tighten anything. The boxes either retain or grow their spread as stitching is added; the means rise modestly, but the scatter rises with them, so the practitioner who removes a bead and adds stitching has bought a higher mean at the cost of a less reliable joint, not a tighter one.

Third, the visible width of the four high-variance boxes (Samples 1, 4, 5, 6) is similar; the two narrow boxes (Samples 2 and 3) are clearly distinct. This is the empirical signature of the weakest-link-to-mean transformation argued in Section 10.3 and Footnote 1, with one important qualifier (Section A.7): in this test, the dominant scatter is substrate variation rather than bond variation, because the bond did not separate in any specimen. The variance reduction of Sample 2 is therefore the signature of a load-redistribution effect through a redundant substrate, not of a bond-strengthening effect through redundant adhesive. The interpretation is the same; the mechanism is one step further upstream than the body text of the chapter has implied.

Fourth, the means are not the whole story. A buyer who is asked to choose between Sample 6 ("save 50 per cent of the adhesive on a two-bead pattern") and Sample 2 ("save 30 per cent of the adhesive on a three-bead pattern") is being asked, in the language of this figure, to choose between a wide grey box at 11 965 N/m and a narrow white box at 16 396 N/m, the latter inside the dashed frame. The two are nominally adjacent on a saving spreadsheet; on the boxplot, they are in different rooms.

A.7 Limitations

Four limitations of the data should be acknowledged. First, the sample size of ten specimens per condition is adequate for the comparison made here — the differences between conditions are large relative to the within-condition scatter — but a larger n would tighten the confidence intervals and would in particular allow a formal test for the variance-reduction effect.

Second, and most consequentially, the test was conducted on B-flute. B-flute is the most common flute in commercial practice and is a representative specimen of the regime in which secondary packaging is mostly produced, so the result is directly applicable to a large fraction of the field. It does not, however, license a generalisation upward. The PTS report concludes that, on this substrate, the bond is stronger than the material, as it did not separate in any of the 60 tests. On heavier flutes — C, double-wall BC and BA, combined-wall EB and AB — the substrate offers more material to resist the FES test, and the test could plausibly recover the bond as the limiting weak point rather than the substrate. The expectation is that flute height grows substrate strength faster than it grows glue-seam strength, so the relative position of bond and substrate on the failure ladder may invert at the upper end. The experiment to confirm that has not been published; the B-flute result should not be extrapolated to AA or BC without testing.

Third, the test was conducted at one folder-gluer line speed of 200 m/min. Higher line speeds may shift the satellite count and the bead-end termination quality — both of which Section 10.4 has identified as practical constraints — and the segmentation pattern itself is harder to print cleanly at the upper end of the speed range with a low-frequency valve. The mean and variance numbers above are therefore best read as the result obtainable on a modern valve at 200 m/min on B-flute, which is one specific corner of a multi-dimensional production envelope.

Fourth, the per-specimen raw values are not in the published report. The five-number summaries in Table 10.A.1 are computed from the published mean and standard deviation under a normal-distribution approximation. Where the bond did not separate, and the substrate failed instead, the empirical distribution will skew slightly toward the lower tail (a low-strength specimen finds substrate weakness earlier), so the actual minimum and Q1 may sit slightly below the values computed here, and the actual maximum slightly closer to the median than computed. The means, standard deviations and CV ratios are exact as published; the quartiles are reasonable.

Chapter 11:

Waterscoring: Treating the Cause, Not the Symptom

Introduction card

Waterscoring reduces the spring-back force that drives many manufacturer's-joint failures. This chapter explains how local crease-line conditioning changes fold mechanics, supports downgauging, reduces adhesive demand, and prevents liner and print-side fractures. It is central for readers evaluating technical interventions on stiff, recycled, lightweight, or digitally printed grades.

"Can the spring-back force at the manufacturer's joint be reduced enough to make stitching, dotting, and narrow tabs viable on stiff and recycled grades, without compromising BCT or pack integrity?"

Economic impact: Waterscoring can reduce adhesive consumption by 30 to 50 per cent per pack. The waterscoring fluid is roughly one tenth of the per-kilogram cost of PVAc dispersion. The saving can combine with stitching, creating a structural reduction in adhesive demand rather than a marginal adjustment.

Ecological impact: Waterscoring supports PPWR-compliant downgauging by preserving crease integrity on lighter boards. Without crease conditioning, downgauged boards are more likely to fracture at the fold. With waterscoring, lighter grades can remain foldable and functional. The ecological benefit comes from less corrugated material, fewer rejected packs, and less embedded CO_2 per delivered functional unit.

Operational efficiency: Fishtailing and gap deviation can cease to be systematic defects when the fold localises reliably at the intended crease. The wetted band becomes the weakest cross-section, so the fold no longer searches for another hinge. Waterscoring also narrows the stiffness difference between flute-tip and flute-valley positions, reducing phase-related variation along the crease.

Quality: The same mechanism that prevents inside-liner secondary fractures also prevents fine white cracks on digitally printed outer liners. At fibre level, moisture temporarily increases ductility. This allows the liner to fold without brittle fracture, which is especially important for continuous print designs crossing the fold.

Regulatory context: Non-ionic surfactants used in commercial waterscoring fluids sit within the EU 1935/2004 framework and BfR Recommendation XXXVI for food-contact paper and board. No additional food-contact authorisation is required beyond the substrate liner.

Process: On coated outer liners, the default application strategy is from the inside of the blank. The applicator is usually mounted below the conveyor. The surfactant chemistry allows fast absorption into the fibre network, which pure water cannot achieve in the same way.

Limitations: Waterscoring is not suitable for barrier-coated grades on the coated face, or where both faces are coated. In these cases, the crease must be solved mechanically through score geometry, bead width, and rule design. The factor-of-three-to-four spring-back reduction is supported across fibre, liner, and board scales, but direct transient production measurements remain limited.

Read in full: production engineers, plant managers planning capex, quality managers building substrate-extension cases, print buyers diagnosing print-side fold-edge defects.

Card sufficient: procurement teams, marketing teams, and executives considering a feasibility study.

Prerequisite: Chapter 8, the secondary fracture and its diagnosis.

Follow-up: Chapter 12, waterscoring at the packing line; Chapter 13, interaction with stitching.

11.1 What Is Waterscoring?

Waterscoring is the targeted application of an aqueous solution to the crease line of corrugated packaging, immediately before or after the creasing operation at the folder-gluer. The solution is water plus a non-ionic surfactant that lowers the surface tension from approximately 72 mN/m (pure water) to approximately 28 mN/m. The fluid is applied through nozzles or applicator wheels, precisely aligned to the crease-line positions.

The principle is simple. Moistening cellulose fibres makes them more pliable. Softening the fibres at the crease before the tool deforms them creates a more precise fold edge, reduces fibre damage, lowers spring-back force and eliminates linear fracture at the fold. The execution is more complex: the moisture quantity must be matched to the liner type (Cobb value), the penetration time must correspond to the machine speed, and the moisture must not migrate into areas that need to remain dry, in particular, the bonding zone of the manufacturer's joint.

A precondition should be applied to every implementation and stated up front. This condition limits where waterscoring can be used: the fluid only works if it can enter the substrate. An open liner, such as uncoated kraft, unsized testliner, or recycled liner, accepts the fluid through the capillary structure of the fibre network. Section 11.3 develops this in detail. A closed surface, for example, a high-density barrier coating, a polymer film, a clay-coated white-top with sufficient hold-out, or a print finish that has filled the surface porosity, refuses entry. No dosage adjustment recovers the effect. The default application strategy on a coated outer liner is therefore to treat the inside of the blank, rather than the printed outside. The inside liner is rarely coated for any purpose other than appearance, and that appearance is rarely a concern in transport packaging. Where both faces are closed (barrier-coated grades discussed in Section 11.6), waterscoring is not applicable to the grade at all. In such cases, the crease must be solved mechanically.

11.2 Waterscoring: Local Action, Not Wetting

Of all the process interventions in this book, waterscoring provokes the strongest instinctive resistance. The objection is almost always the same and is usually voiced before any technical detail has been heard: water is the enemy of corrugated board. Liners soften, flutes collapse, BCT drops, stacks fail. Against that background, the idea of deliberately applying water to a corrugated blank sounds counterintuitive at best and negligent at worst. The objection is legitimate; it is also misdirected, because waterscoring is not the same as wetting a corrugated blank. It is a strictly localised, transient plasticisation of the crease line alone, which leaves the structural board untouched.

What waterscoring actually does

Waterscoring deposits a controlled amount of water — typically a few grams per square metre — onto a narrow band along the intended fold. Two configurations are in use. In the retrospective configuration, fluid is applied to an already-impressed crease shortly before folding; the water migrates into the compressed fibre network of the score and softens the cellulose-hemicellulose-lignin matrix, where plastic deformation must occur. In the process-integrated configuration, the fluid is applied to the intended crease path immediately before the rule engages; the score is then impressed into a locally plasticised line, and the fold follows directly. Either way, the essential property is locality. The applicator addresses a band typically two to four millimetres wide, centred on the crease. The surrounding panel, the flutes, and the inner and outer liners outside this band receive no water. The mass of fluid per blank runs from tens to a few hundred milligrams, depending on box size and score length — a per-mille moisture increment of the blank, confined to a line.

Why does the operator see nothing?

An operator watching the line sees no wet stripes leaving the machine. This surprises visitors and reassures customers. The reason is twofold. Corrugated board is an extremely effective wick: the fibres, the starch bonds and the capillary structure of the liner absorb and redistribute water within seconds. A moisture front that was a visible line at the moment of application has, by the time the blank reaches the folding station, diffused laterally and into the depth of the liner enough that no optical contrast remains. The fluid is formulated to assist this redistribution. Plain water on a sized liner beads up; the surfactant lowers the surface tension below the liner's critical surface energy, and the fluid spreads instead, drawn into the capillary structure rather than sitting on its surface. What would be a visible droplet becomes, in the same instant, a sub-surface moisture distribution.

There is a simple in-line check. Wipe a cloth moistened with clean water across the area where the waterscore was applied; a contrast line briefly appears, because the waterscored band, already at slightly elevated moisture, accepts the additional water differently from the surrounding dry liner. The contrast fades within a minute as equilibrium is re-established. The presence of waterscoring is verified not by observing the blank itself but by revealing it with a second controlled wetting. The day-to-day result is that packs leaving the line look, feel and handle like dry packs — because outside the few square centimetres of the score they are dry.

BCT and the structural board

Customers new to waterscoring almost invariably ask for a Box Compression Test comparison, almost invariably measured immediately after production, on the assumption that residual moisture will be at its worst at this point and depress BCT. The result, repeatedly confirmed in field trials, is that BCT measured directly after production on a waterscored pack is statistically indistinguishable from BCT of the same design produced dry. BCT is dominated by panel buckling and the edge support of the vertical scores. The horizontal score that is being waterscored at the packing line contributes negligibly to the buckling-relevant area; the vertical scores are not waterscored at all. The integrated moisture across the pack stays within normal production and storage variability, and the load-bearing structure — the flutes and the unscored liner regions — is untouched.

This is not a theoretical claim; it is the findings customers report from their own incoming-goods testing under their own protocols. It is also the finding that lets waterscoring be introduced without renegotiating BCT specifications: the specification does not change because the measured property does not change. What does change is the mechanics of the crease itself, and that change is large — the subject of Sections 11.3 and 11.4.

Engineering parameters and what waterscoring is not

Implementation is not indifferent to detail. The applicator must be metered (too little water fails to plasticise; too much wets the surrounding liner, marks the surface, and at the extreme reduces local strength). The fluid must be clean, because contaminants concentrate at the score and can interact with print and coating. The dose must be adjusted for board grade and flute, because absorption differs (Section 11.6). The interval between application and folding must be controlled, as plasticisation has a characteristic window of optimal action. None of these parameters, however, addresses the customer objection of the opening paragraph. None of them produces a wet pack, depresses BCT, or introduces moisture into the structural board. They determine whether waterscoring works well or works poorly; they do not determine whether waterscoring compromises the pack, because at the quantities involved, it does not. The right mental model is not "water applied to corrugated board" but "plasticiser applied to a hinge". The plasticiser is water, which is cheap, food-safe, traceless, and reversible; its function is that of a local, transient process aid, not a bulk treatment.

11.3 Why Water Alone Is Not Enough

Figure 36 Pure water (top) and waterscoring fluid (bottom) on a kraftliner surface. Both volumes are deliberately well above the production dosage, so the wetting behaviour is visible to the eye. Pure water sits as spherical caps after several seconds — high contact angle, near 85°, beading rather than spreading. The surfactant solution drops the contact angle to roughly 30°, spreads instantly into a thin film and is wicked into the fibre network within microseconds, leaving only the wetted shadow visible. This is the practical illustration of the Washburn argument in Section 11.3 — same fluid mass, opposite outcome.

Pure water at 20 °C has a surface tension of 72 mN/m. On a sized liner, it forms a high-contact-angle dome; the droplet sits, runs off, and only slowly penetrates. Capillary forces are insufficient to draw the liquid into the fibre network during the available contact time, which at production speed is fractions of a second.

The basic physics is the Washburn equation [81], which describes capillary liquid transport into porous media:

$$L^2 = (\gamma \cdot \cos\theta \cdot r \cdot t) / (2\eta)$$

Here, L is the penetration depth, γ the surface tension, θ the contact angle of the liquid against the substrate, r the effective capillary radius, t the time and η the dynamic viscosity. The product γ·cosθ is the relevant grouping: it represents the net driving force for capillary flow after accounting for the wetting condition.

Plug in the numbers. Pure water on an untreated-sized kraftliner has a contact angle of 80 to 90 degrees; using cos(85°) ≈ 0.087, the product γ·cosθ is approximately 6.3 mN/m. The water-scrubbing fluid has a surface tension of 28 mN/m and a contact angle of 20–40°; using cos(30°) ≈ 0.866, the product γ·cosθ is approximately 24.2 mN/m. The ratio of the two products is 24.2 / 6.3 ≈ 3.9. That single number drives every other comparison.

What "3.9" means depends on which question is asked. At equal contact time, penetration depth scales with the square root of γ·cosθ: the surfactant fluid penetrates to a depth that is √3.9 ≈ 2.0 times greater than pure water, so the depth advantage is approximately 2×. At equal penetration depth, however, the time required scales linearly with the inverse of γ·cosθ: the time advantage — i.e. how much faster the surfactant fluid reaches a given depth — is the full 3.9, so approximately 4×. The same holds for the volume of fluid taken into the liner per unit area in a given time, which scales with depth and is therefore approximately 4× (and, under the more extreme contact angles of heavily sized kraftliner, where pure water approaches 88 to 89 degrees, can rise toward 6×). All three numbers — the 2× depth ratio at equal time, the 4× time-to-equal-depth ratio, and the 6× ceiling under highly sized conditions — are the same physical effect read on different axes; quoting any one of them in isolation invites the suspicion of contradiction that earlier drafts of this book accidentally produced. The conservative, geometry-independent number is the depth ratio of two, and that is the number used in the rest of the chapter.

The practical consequence holds either way. Moistening the crease line with pure water wastes the greater part of the fluid: it sits on the surface, migrates laterally, and does not reach the deformation zone in time. The surfactant solution penetrates the capillary structure immediately and reaches the fibres that are about to be deformed by the rule.

11.4 Mode of Action: How Waterscoring Addresses the Root Causes

Waterscoring operates through three interconnected mechanisms, each tackling a problem documented in earlier chapters.

Mechanism 1: Fishtailing and gap deviation

Fishtailing — the non-parallel closure of the manufacturer's joint, in which the top and bottom of the glue tab meet the opposing panel at different positions — and gap deviation — the resulting wedge-shaped gap that runs along the pack height, tight at the top and open at the bottom or the other way around — are two readings of the same root cause. The fold finds the weakest cross-section it can find. In dry corrugated board, the crease line is only one weak cross-section among several. Flute valleys, adhesive lines, localised fibre damage from the corrugator and pre-crushed zones from the conveying nip all compete with the score. Where any of these is locally weaker than the score, the board hinges there instead. The fold drifts laterally relative to the intended line; one side of the panel ends up with a longer hinge arm than the other; the joint closes unequally at the top and the bottom. The lateral drift is what the operator calls fishtailing; the vertical asymmetry is what the case erector reads as gap deviation. They are the same defect, viewed from two directions.

Waterscoring acts on this through two complementary effects. The first is localisation. The applicator deposits a fluid strip two to four millimetres wide, centred on the score; that band plasticises within seconds before folding. The score is then unambiguously the weakest cross-section in the panel — not by a marginal stiffness difference that the corrugator's small irregularities can override, but by the factor of three to four documented in Section 11.4 (Mechanism 2). The fold localises at the wetted band with no competition from the surrounding dry material. The condition for this argument to hold is that the moisture line sits exactly above the rule line. In production, this is solved geometrically rather than by re-aiming the applicator for each job: the valves of the waterscoring head are mounted on the same width-adjustable folding bar as the rule, so any lateral repositioning of the rule for a different blank format carries the moisture line with it. The applicator follows the bar; the bar follows the format. Misalignment of the moisture line is therefore a setup error rather than a running drift, and a properly mounted system does not require re-tracking during a run.

The second effect is harmonisation, and it is the subject of Section 11.5 in full — worth previewing here because it is the missing half of the answer to why fishtailing disappears. The dry score shows unequal stiffness along its length, depending on which flute position the rule engages at each point. At the apex, the rule has the highest dry stiffness and the fold demands the highest moment; at the valley, the rule meets the lowest dry stiffness and the fold demands the lowest moment; on the shoulder, the contact is asymmetric, and one side of the score deforms more than the other. These three regimes alternate along every transverse score, modulating the fold mechanics from one flute pitch to the next. The lateral force distribution along the score follows the modulation, and the unequal force pattern is what tilts the fold off the intended axis and produces the wedge-shaped gap. Waterscoring narrows the spread between the three regimes — most strongly at the apex, where the moisture has the most material to plasticise, and least strongly at the valley, where the dry stiffness is already low. The geometric asymmetry remains, but the difference in resulting force is compressed by a factor large enough that the lateral forces along the score no longer drive the fold off-axis.

Both effects converge on the same outcome. Localisation removes the competition for the fold; harmonisation removes the lateral forces that would still tilt the fold even after localisation. Fishtailing disappears as a systematic defect because the score is now the only place the fold can go, and gap deviation collapses because the force distribution along the score is no longer modulated by the flute phase the rule happens to be in. The two defects share a mechanical signature; waterscoring is removed by a single intervention applied to a few square centimetres of board per pack, and the fuller treatment of the harmonisation half is found in Section 11.5.

Mechanism 2: Reduction of spring-back forces

Chapter 3 identified spring-back force as the primary driver of MJ failure: the bending stiffness at the crease line generates an opening moment that acts on the still-fresh glue line as peel or cleavage. The stiffer the board at the score, the higher the spring-back force, and the larger the load the green PVAc bond has to resist before it has set. Bending stiffness scales with the cube of material thickness and is proportional to the elastic modulus. Waterscoring reduces the local elastic modulus at the crease line by a factor of three to four, and the spring-back force falls in the same proportion. This shifts the ratio of spring-back force to glue-line strength out of the critical range during the first minutes after bonding, when the dispersion has reached only five to fifteen per cent of its final strength.

The factor-of-three-to-four claim deserves an explicit evidence base, because the most direct primary measurement is from a different scenario. Mrówczyński et al. [82] measured the mechanical properties of corrugated board across a matrix of temperature and ambient humidity steps. At 90 per cent ambient relative humidity, bending stiffness drops to thirty-three to fifty-five per cent of the dry-reference value depending on flute and direction; torsional stiffness falls to about thirty-five per cent; in-plane shear stiffness to as low as twenty-eight per cent. Above 70% RH, the curve steepens sharply, and by 90%, the fibre network has lost most of its rigidity. These are bulk-equilibrium values measured after the entire specimen has equilibrated to the chamber's elevated humidity.

Waterscoring is a different scenario in three respects. The moisture is local rather than bulk, applied to a band of two to four millimetres rather than the whole specimen. The moisture is transient rather than equilibrium-bound, present at the crease only for the seconds spanning deformation and bond formation. And the moisture is delivered as a metered liquid film with a surfactant, not as ambient water vapour, so the local moisture content at the score during plasticisation reaches values that bulk equilibration at any feasible chamber RH cannot match. The transferability argument is therefore not that waterscoring reproduces a 90 per cent RH chamber. It is that the waterscoring scenario delivers, at the score, more moisture for less time than the chamber delivers at the whole specimen, and the same molecular mechanism is responsible for the stiffness drop in both.

That mechanism is documented down to the single-fibre level. Auernhammer et al. [104] mapped humidity-dependent mechanical properties along individual cellulose fibres using atomic force microscopy, recording the local stress-strain response, contact stiffness, and dissipation as the ambient moisture varied. Their measurement is the cleanest available evidence that the fibre's elastic modulus drops with moisture on a time scale set by hydrogen-bond exchange (milliseconds), independently of any bulk-equilibration argument. Vonk et al. [17, 80] showed in parallel that the same fibres swell transversely by twenty to thirty per cent on moisture absorption, with the hygro-expansion mapped at full-field resolution; the hydrogen bonds between fibrils are temporarily replaced by water molecules, and the fibre transitions from elastic-brittle to ductile within milliseconds of wetting. The molecular mechanism does not require equilibration at the panel scale to act — it acts at the fibre, in the time it takes for water to reach the fibrils.

Several macroscopic measurements anchor the same effect at the panel scale. Navaranjan et al. [106] tested both virgin and recycled corrugated paperboard at 50 per cent and 90 per cent RH and reported elastic-modulus drops of comparable magnitude to Mrówczyński's, with virgin liners outperforming recycled ones at both humidities.

Cornaggia et al. [3] documented a comparable steepening of the stiffness curve above sixty-five per cent RH over various liner combinations. Szewczyk and Głowacki [15] reported the same trend on linear tensile strength and burst, with thirty to fifty per cent drops between fifty and ninety per cent RH.

Modzelewska [10] independently confirmed the climatic-condition-strength relationship for the corrugated case. At the crease itself, Walker and Seffen [105] modelled the flexural mechanics of a creased thin strip, treating the crease as a shallow cylindrical hinge with continuous structural properties; their analysis shows that the bending moment required to fold and the spring-back moment recovered after release scale together with the local elastic modulus of the score, so a stiffness drop at the score translates directly and proportionally into a spring-back drop.

Beex and Peerlings [98], Huang, Hagman and Nygårds [100], and the recent continuum damage model of Robertsson et al. [107] confirm the same one-to-one translation in their respective creasing-specific frameworks: the bending moment under the rule and the resulting spring-back after release scale with the same elastic modulus factor.

Three arguments therefore back the factor-of-three-to-four claim, none resting on Mrówczyński alone.

First, the underlying mechanism is fibre-level hygro-expansion and modulus drop (Auernhammer [104], Vonk [17, 80]), which acts on a millisecond time scale and does not depend on bulk equilibration.

Second, the macroscopic stiffness drop at high local moisture is independently confirmed at the panel scale by Cornaggia [3], Szewczyk and Głowacki [15], Modzelewska [10] and Navaranjan [106], and at the score scale specifically by Walker and Seffen [105], Beex and Peerlings [98], Huang/Hagman/Nygårds [100] and Robertsson [107].

Third, practical experience confirms the same number on the production line. On a folder-gluer where waterscoring can be switched on and off between blanks, the change in fold mechanics is immediate and visible: spring-back force drops to roughly one-third to one-quarter, fishtailing disappears, and the green bond at the MJ stops opening. Photographic before-and-after evidence of the same crease, folded with and without waterscoring, is the standard demonstration material in technical bulletins; it shows the same effect each time across all board grades.

The factor of three to four is not a single point in the literature. It is a mechanism (fibre hygro-expansion) confirmed at three scales (fibre, liner, board) and visible at the fourth (the produced pack), with the chamber-RH measurements bounding the local-moisture state from below — not from above, because the score sees, transiently, more moisture than any feasible chamber RH can deliver at equilibrium.

Mechanism 3: Increased liner elasticity

The third mechanism concerns the liner itself. Cellulose fibres are hygroscopic: on moisture absorption, they swell transversely by twenty to thirty per cent. The hydrogen bonds between fibrils are temporarily replaced by water molecules, and the liner becomes ductile rather than brittle. Without waterscoring, the liner tears at the fold edge, particularly in kraft liner with high tensile strength but low elongation at break and in lightweight grades with less fibre reserve. A linear fracture creates a crack that acts as a stress concentrator and can propagate under load. Adjacent to the MJ, such a crack is a joint weakness that no adhesive can compensate for. Waterscoring eliminates the fracture initiation: the softened fibres deform plastically rather than fracturing. No linear fracture, no crack-initiation point next to the MJ.

This claim is testable from the material side. The IGF 21804 BR sensitivity analysis [103], which correlated 159 measured material parameters with crease quality across 11 corrugated grades, found no single material parameter to be dominant in the diagnosis but identified high elongation at break as the qualitative signature of stable creasing. Waterscoring is the process intervention that converts a low-elongation kraft liner into a high-elongation surface, locally and reversibly, exactly at the moment the rule strikes. The match between the diagnostic property of Section 8.5 and the therapeutic mechanism of this section is a single variable, addressed by a single intervention. The chapter that defines the disease names the cure.

The same elasticity gain is visible on both the print and fold sides. Without waterscoring, the outer liner that carries the print fractures at the fold in the same way the inside liner does, but on the printed face, the fracture is visually amplified: a fine white hairline runs along the fold edge wherever the fibre network has separated and exposed the unprinted substructure. This is the defect print buyers spot first, especially on dark backgrounds and saturated colours, and it is the defect that loses premium-print orders.

Digital-printed corrugated is particularly vulnerable because the inkjet design typically extends across the entire pack, including the fold lines , the continuous-print aesthetic that is one of the digital-print sales arguments, so every crease becomes a potential print defect. At the score itself, where the rule has locally collapsed the flute structure and brought the inside and outside liners into close contact, fluid applied from the inside reaches the outside fibre network through the compressed cross-section before folding.

The outer liner, therefore, plasticises at the fold even when the application is on the inside, and the print continues across the fold without the white-crack signature. The argument that the inside liner does not crack and that the printed outside does not show white-crack marks is the same argument applied to the two faces of the same fibre network.

11.5 Crease Position and Flute Geometry

This section develops the harmonisation half of Mechanism 1 (Section 11.4). The localisation half is well-defined by the geometry of the applicator and its mounting on the folding bar; the harmonisation half deserves its own treatment because the force-modulation effect along the score is the part of the fishtailing-and-gap-deviation argument that is otherwise easy to miss.

Chapter 8.4 identified the flute-phase position as one of the most significant and uncontrollable sources of variation, with the IGF 21804 BR finite-element investigation [103] adding the position-of-crease distinction (wave-crest, valley, intermediate). Nagasawa and Komiyama [29] quantified the relationship for transverse creases: the position of the crease relative to the flute geometry influences the spring-back moment according to $\Delta M\_peak = 2.21 \times (e/\lambda) + 0.34$, where e is the eccentricity of the crease relative to the flute valley and λ the flute wavelength. Nica et al. [30] confirmed the same picture for inline creasing at the corrugator. The result of either configuration is uncontrollable scatter in the increase in quality along each crease line.

Three contact positions are relevant on a transverse crease. At the apex, the rule meets maximum resistance because the flute tip is the stiffest point in the structure; the creasing force is highest, and the risk of linear fracture is at its maximum. On the flute shoulder, the rule meets the structure asymmetrically; one side of the score is deformed more than the other, producing an asymmetric crease with one-sided force introduction during folding. In the valley, the rule meets minimum resistance; the score is formed with the lowest force and the least fibre damage, but at the cost of excessive crease depth, which can reduce panel stiffness. The dry crease, therefore, varies in quality along its length according to which of the three positions the rule happens to engage at each point.

Waterscoring homogenises the result. Fibre softening reduces the bending-stiffness difference between the three positions: at the apex, where the force is highest, the softening effect is greatest because the moisture has the most material to plasticise; in the valley, where the force is already low, the softening matters less. The geometric asymmetry remains — the rule still meets a tip in one place and a valley in another — but the difference in resulting crease quality is compressed by a factor that significantly homogenises the fold along the entire crease. Since BCT under stack loading is more sensitive to the weakest crease than to the average, this homogenisation is an underestimated contribution to BCT in real pallet configurations, even when the laboratory BCT is unchanged.

Read alongside Mechanism 1 of Section 11.4: localisation makes the score the only place the fold can go; harmonisation removes the modulation of force along the score that would otherwise tilt the fold off-axis even after it had localised. The two together explain why fishtailing and gap deviation are not simply reduced by waterscoring — they cease to be systematic defects, because the two mechanical conditions that produced them have both been neutralised.

11.6 Fitting the Dose to the Liner

The correct dosage depends on the liner's absorption behaviour. The governing parameter is the $Cobb_{60}$ value, as specified in ISO 535 [75], which measures water absorption in g/m^2 after 60 seconds. Kraft liner with low $Cobb_{60}$ needs more fluid and a longer contact time to achieve the desired softening; recycled liner with high $Cobb_{60}$ absorbs the fluid immediately, but the risk of strike-through to the opposite surface rises in step. Dosage is therefore matched not only to liner type but also to machine speed and the distance between the application point and the rule.

$Cobb_{60}$ values and waterscoring dosage strategy by liner type.

Kraftliner (sized top liner)	25 to 35	Slow; high water resistance	Higher dosage, longer contact time
Testliner (sized)	30 to 45	Moderate; batch-dependent	Standard dosage
Testliner (unsized)	60 to 90	Fast; risk of strike-through	Reduced dosage
Recycled liner	80 to 120	Very fast; high strike-through risk	Minimum dosage; monitor closely
White-top liner	25 to 40	Moderate to slow; coating-dependent	Standard; caution with coated variants

Two cases lie entirely outside the dosage matrix. The first is barrier-coated corrugated and carton grades that carry a polymer or clay barrier on at least one face for grease, moisture, oxygen or water-vapour resistance. The barrier coating's defining property is water repellency, which is precisely what makes it useful in its intended application.

A fluid applied to the coated face cannot penetrate it, and no surfactant addition or dosage increase restores the effect; the coating's barrier function is also its criterion for water-scoring exclusion. Waterscoring is therefore not applicable to barrier-coated grades on the coated face. If both faces are coated, waterscoring is not applicable to the grade at all, and the crease-stiffness reduction must be achieved solely by mechanical means, such as a deeper score, a narrower bead, or a multi-rule construction matched to the fold geometry. Plant managers running mixed liner streams should treat barrier-coated lines as a separate process class.

The second case is the application-side question. The default strategy on grades with a coated or closed outer liner, including coated white-top liner, fully sealed UV-printed surfaces and high-coverage overprint-varnished blanks, is to apply the conditioning fluid to the inside face of the blank rather than the printed outside. The inside liner is rarely coated for functional reasons (the print is on the outside, the food-contact rating is in the outside coating, the barrier in the outside polymer), and the inside therefore accepts the fluid with the standard dosage matched to its own $Cobb_{60}$.

At the score itself, where the rule has compressed the flute structure, and the two liners are in close contact, fluid applied from the inside reaches the outside fibre network through the collapsed cross-section in time for the fold, and the elasticity benefit is delivered to both faces. Inside application is the operationally simpler default in commercial installations, regardless of whether the outside is coated, because it also removes the print-fixation timing question that arises when conditioning fluid and freshly-printed ink would otherwise meet on the same face (Section 12.7 develops this for the packing-line case).

A practical consequence of the inside-application default is that the applicator typically sprays upward, against gravity, onto the underside of the blank as it passes over the conveyor. The first reaction of the operator who has not seen this is that the fluid will drip back. In commercial installations under normal dosing, it does not, and the physics is the same surfactant chemistry that makes waterscoring work in the first place. Pure water on a sized liner beads at a contact angle of 80 to 90 degrees; as deposition continues, the bead grows past the capillary length (about 2.7 mm for water under gravity), and a droplet detaches and falls. For that reason alone, pure water cannot be applied from below.

The surfactant solution at 28 mN/m drops the contact angle into the 20-to-40-degree range; the fluid spreads as a thin film of roughly 10 μm thickness for a 10 g/m^2 dose, so the entire deposited volume sits in capillary contact with the fibre network at once. The Washburn equation in Section 11.3 predicts the wicking time for that film thickness to be on the order of microseconds (about 8 μs for 10 μm of penetration at typical dosing), three to four orders of magnitude faster than any gravitational drip mechanism can act.

By the time a hanging droplet would form, the fluid is already inside the liner. The geometry is permitted by the surfactant; without it, application from below is not feasible. The observation that no dripping is seen on a properly dosed from-below applicator is therefore not an empirical curiosity but a direct consequence of the surface-tension reduction the chapter has already used to explain wetting and penetration. Same physics, different axis.

11.7 Waterscoring and the Wider System

Waterscoring interacts with most of the mechanisms described in Chapters 2–10. The interactions are summarised in Table 11.2 and discussed in turn below.

Waterscoring interactions with the wider production system.

Washboarding (Chapter 2)	At amplitudes above ~ 50 µm, fluid drains into the valleys; surface contact at the peak is lost.	Surfactant solution wets both peaks and valleys; the effect is reduced on heavy washboarding but not eliminated.
Warp (Chapter 6)	Spring-back energy from the warped panel acts on the MJ as peel; the joint is loaded before it has set.	Stiffness reduction at the crease lowers spring-back proportionally; the warp is unchanged but the load it produces drops.
Stitching (Chapter 10)	Each segment must carry the full local peel force; the redundancy is structural, not energetic.	Each segment carries less; the same redundancy now sits in front of a smaller load. The two interventions multiply rather than add.
Narrow tabs / PPWR (Chapter 9)	Tighter clearances against the crease and higher peel-stress concentration; the dry crease imposes more	Lower spring-back means lower peel stress; tabs that are unreliable when dry-creased work correctly with waterscoring.

	force on a narrower geometry.	
Liner type (Section 11.6)	Effect dominated by liner properties; brittle kraftliner cracks at the crease, lightweight liners tear, recycled liners variable.	Effective on all liner types with a dose matched to $Cobb_{60}$; the effect equalises crease quality across the liner range.

Each row of the table reflects a chapter that has already been read. The washboarding interaction (Chapter 2) is the only one in which waterscoring is partially constrained by an upstream defect: when the surface waveform exceeds roughly fifty micrometres in amplitude, fluid drains into the valleys, and the peak fibres receive less. The effect is reduced but not eliminated, because the score itself is a valley in the topography — the rule has already created a depression at the deformation site — and the surfactant solution still wets the fibres there. The warp interaction (Chapter 6) is unambiguously positive: warp produces spring-back, waterscoring reduces spring-back, the warp is unchanged, but the load it imposes drops by the stiffness factor of three to four.

The stitching interaction (Chapter 10) is where the two interventions become each other's force multiplier. Stitching makes the joint damage-tolerant by fragmenting the crack path; waterscoring lowers the load that any one segment must hold. The two effects multiply rather than add: a 30-per-cent-stitching pattern that would deliver a +75 per cent gain in Factory Edge Strength against a dry continuous bead does so against a glue line that is also exposed to a smaller spring-back load. The narrow-tab interaction (PPWR-driven, Chapter 9) is similar in shape: the tighter geometry imposes higher peel stress on the joint, and waterscoring brings that stress back into the regime where the joint can still hold. Tabs that are unreliable when dry-creased work correctly with waterscoring; the PPWR push toward narrower tabs is feasible only because waterscoring exists. The liner-type interaction (Section 11.6) closes the loop: the same dose-matching that makes waterscoring work across the liner range also equalises crease quality across that range, so the box engineer is no longer locked into a single liner family by the crease.

11.8 Less Material, Better Boxes: Waterscoring and the Light-Weight Imperative

Lighter paper weights (downgauging) are one of the most effective levers for PPWR compliance. Chapter 2 showed that downgauging amplifies washboarding, increases moisture sensitivity and narrows tolerance windows; lighter liners are more brittle, thinner, and more prone to fracture at the crease. Waterscoring resolves this directly. It enables crack-free folding of lightweight boards that could no longer be creased reliably when dry. A 100 g/m^2 testliner that tears at the crease when dry folds cleanly with waterscoring; a BC double-wall board with 120 g/m^2 facings that generates a dry spring-back moment exceeding the MJ operates within the safe range when the score is plasticised.

The integrated concept

Waterscoring, stitching, and narrow tabs combine into a coherent answer to PPWR. Less paper through downgauging — enabled by waterscoring, which keeps lighter boards creasable. Less adhesive through stitching, which can cut consumption by up to half by shifting the design philosophy from maximum strength to damage tolerance. More precise folding from waterscoring, which eliminates fishtailing and tightens gap variation. More reliable joints from the combination, in which a lower spring-back load meets a damage-tolerant glue pattern. The four elements are interdependent: downgauging without waterscoring leads to liner fractures; stitching without waterscoring increases fishtailing; narrower tabs without either intervention raise peel stress beyond the bond's capacity. Only in combination does the system become simultaneously lighter, more material-efficient and more reliable.

11.9 Long-Term Behaviour and Regulatory Status

Long-term behaviour

Two long-term concerns are sometimes raised by customers and deserve a direct answer. The first is microbial growth: can the transiently elevated moisture at the crease line, sustained over weeks of warehouse storage, support mould or yeast development at the score? In practice, this does not occur for the same locality reason that BCT is unaffected. The deposited mass is in the milligram range, the fluid distributes laterally and into the liner depth within seconds of application, and by the time the pack reaches the warehouse, there is no concentrated moisture pocket at the score for an organism to colonise. A pack opened weeks after production shows no visible moisture stripe at the score; the only way to make the score visible again is to wipe the area with a damp cloth, which produces a brief contrast line as the slightly elevated equilibrium moisture of the score accepts additional water differently from the surrounding liner. That contrast itself demonstrates that the score is in moisture equilibrium with the rest of the panel, not above it. The neighbouring observation from Csavajda and Böröcz [108], who measured BCT recovery in corrugated boxes after intentional moisture exposure, is consistent with this picture at much higher dosage levels: even after substantial wetting and re-equilibration, the BCT recovery is largely complete once the moisture redistributes, which means the small per-mille moisture deposit of waterscoring leaves no measurable residue at warehouse time scales.

The second concern is interaction with print and coating. Ink migration, surface tackiness, and odour development at the score have been raised as theoretical risks. Practical experience with thousands of installed lines indicates that they do not deliver the proper metered dosage. Where they have been reported, the cause has consistently been overdosing (a dosage tuned for a high-Cobb recycled liner and then applied unchanged to a low-Cobb-sized kraftliner) or contamination of the fluid reservoir, neither of which is a property of waterscoring as such. A clean, properly metered installation produces no detectable interaction with the print or the coating, no detectable odour at the score, and no measurable change in surface tack.

Regulatory status

Where the pack is destined for food contact, the regulatory framework is EU Framework Regulation 1935/2004 on materials and articles intended to come into contact with food, supplemented at the German national level by the BfR (Bundesinstitut für Risikobewertung) Recommendation XXXVI on paper and board for food contact. The non-ionic surfactants used in commercial waterscoring fluids are within the scope of these recommendations and are approved for use in food-contact paper and board production; the supplier's product data sheet documents compliance, and the use of waterscoring does not require an additional food-contact authorisation beyond what the substrate liner already needs. The carrier is water; the additive load is in the per-mille range of the deposited fluid, which is itself in the per-mille range of the blank mass; the integrated additive load on the finished pack is several orders of magnitude below any limit that the food-contact frame imposes. Waterscoring is a regulated, approved process for food-contact packaging when used with the standard commercial fluid. Plants using non-standard fluids should obtain explicit confirmation from the supplier.

11.10 Summary

Waterscoring is a strictly localised, transient plasticisation of the crease line. The deposited fluid is water with a non-ionic surfactant, in a per-mille moisture increment confined to a band two to four millimetres wide. The board absorbs and redistributes the fluid within seconds; the operator sees no wet stripe; BCT measured immediately after production is statistically unchanged from the dry reference.

What changes is the mechanics of the crease. Bending stiffness at the score falls by a factor of three to four — a number anchored at three independent scales (single-fibre hygro-expansion, liner-level humidity studies, paperboard creasing experiments) and confirmed in production by the immediate visible change in fold mechanics when the system is switched on. Spring-back force falls in the same ratio, peel loading at the green MJ drops below the cohesion threshold of the setting PVAc, fishtailing localises away, gap variation collapses, and stack geometry becomes reproducible.

Waterscoring multiplies with stitching (Chapter 10) and with narrow tabs (PPWR, Chapter 9): each of those interventions sees a smaller load applied to a more redundant geometry. It enables downgauging without losing crease integrity. Long-term storage is unaffected by the deposited moisture, which equilibrates with the panel within seconds; mould, ink migration and odour are not observed at properly metered dosage. The non-ionic surfactant is approved for food-contact applications under EU 1935/2004 and BfR Recommendation XXXVI. The intuition that water is the enemy of corrugated board holds in the contexts that produced it; waterscoring is not one of those contexts.

Chapter 12:

Beyond the Folder-Gluer: Waterscoring at the Packing Line

Introduction card

Crease-line conditioning can also pay back after conversion. This chapter examines where waterscoring belongs on packing lines, how it affects hot-melt use and stoppages, and why pack-on-demand throughput figures must be read below nameplate speed. It is relevant for line managers, operations directors, packing-line engineers, and digital-print integration teams.

"Where else in the value chain does crease-line conditioning pay back, why are published throughput numbers for pack-on-demand systems higher than field reality, and what timing and geometry questions arise when conditioning meets digital printing on the same line?"

Economic impact: Packing-line stoppage costs vary strongly by line class. High-volume FMCG lines can reach about EUR 25,000 per hour, while mid-volume FMCG lines often sit around EUR 10,000 to 20,000 per hour and low-volume speciality or contract packing around EUR 3,000 to 8,000 per hour. Stage 1 conditioning treats longitudinal end-panel creases with a single nozzle bar and can reduce hot-melt use and tray reopens. Stage 2 adds transverse conditioning for side-panel creases and unlocks the larger share of spring-back reduction.

Operational efficiency: Crease-quality-related disruptions contribute to the OEE gap, but they are often recorded as "material defects" or "miscellaneous downtime". The chapter links the OEE metric already used by line managers with a crease-quality intervention that converters can supply.

Quality: Published pack-on-demand throughput figures must be read carefully. OEM nameplate values of 1,028 to 1,200 boxes per hour often translate into sustained real-world rates of about 600 to 960 boxes per hour, depending on box mix, print integration, infeed gating, and format changes. Investment cases should be based on the lower sustained bracket, not on ideal nameplate performance.

Process: Fanfold perforations can create competing fold lines between the inline crease and factory-applied perforations. Waterscoring is preferred in this configuration because a conditioned inline score becomes reliably softer than a dry perforation and reasserts itself as the intended hinge.

Digital printing: From-below applicator geometry allows conditioning on the inside of the blank while the print head addresses the outside face from above. The main integration question is timing: each face must complete its prior process step before the next process engages. For digitally printed continuous designs, waterscoring can also prevent the fine white hairline crack that appears only after folding and is often detected at incoming-goods inspection.

Limitations: PSA tape closure is a valid option for hand-close and low-throughput lines, but it is not a substitute for crease conditioning. It resists spring-back rather than reducing it. Laser perforation is also not a documented industrial creasing alternative for corrugated. Its structural damage is permanent, whereas waterscoring creates temporary plasticisation.

Read in full: packing-line engineers, line managers, operations directors, pack-on-demand system specifiers.

Card sufficient: executives evaluating the business case for packing-line intervention.

Prerequisite: Chapter 11.

Follow-up: Chapter 13, the integrated concept.

12.1 From the Corrugated Board Plant to the Packing Line

The preceding chapters have focused on processes inside the corrugated board plant and in converting: washboarding, warp, crease quality, and adhesive application at the folder-gluer. The blank that leaves the plant is, in many cases, not yet a finished pack. It becomes one at the packing line — at the filler, the distributor, the e-commerce shipper. There, blanks are folded around products, erected, glued and closed. The machines that do this impose their own demands on the blank, particularly on the crease lines. A crease that works well at the folder-gluer can cause problems at a wrap-around packer, because the speeds, forces and geometries are not the same.

This chapter analyses the machine types, their requirements for crease quality, the typical failure patterns and the as-yet-untapped potential of crease-line conditioning at the packing line itself.

12.2 Two Philosophies: Pre-Scored Blanks vs Inline Creasing

Three machine concepts dominate the European consumer-goods, beverage, and e-commerce packing lines, each with its own approach to crease lines.

Wrap-around packers and tray formers

The principle is straightforward and efficient. Flat, pre-manufactured, pre-scored blanks are taken from a magazine, the product is placed on the blank, the side panels are folded up and bonded with hot-melt adhesive. The creasing is not done at the packing line; it has already been pressed in at the corrugator or at the die-cutter. The machine does not crease; it folds and glues. Machine speeds run from 15 to 100 blanks per minute, and on beverage trays, the line speed reaches 80,000 to 120,000 cans per hour. The decisive criterion from the blank's side is crease quality — how cleanly the pre-scored line folds, how high the spring-back force is after folding, how reproducible the fold position is across thousands of blanks. Any imprecision is multiplied at these speeds.

Pack-on-demand systems

Pack-on-demand is a different concept. From a continuous corrugated stock (fanfold), a custom-sized pack is cut, creased, folded and closed in real time, all in one machine. The creasing is not advanced work; it is part of the packaging operation itself. The leading systems include Packsize [89], CMC Machinery [90], Sparck Technologies [96], and Kolbus Autobox [97], which predominantly use fanfold corrugated stock in single- and double-wall constructions. Creasing is performed by rotating crease wheels or tooling that engages the material at high speed.

The headline throughput numbers from the OEMs deserve a caveat. The values printed on the data sheets — 1,200 boxes per hour for Packsize and Kolbus, 1,100 for Sparck, 1,028 for CMC — are nameplate maxima, achieved under ideal conditions: a uniform box-size mix, no cycle losses on small boxes, no infeed gating, no print-integration delay, no operator interventions. Realistic sustained throughput in field installations runs 20 to 40 per cent below the nameplate, depending on the box mix, print integration, and the operator's response to format changeovers. Table 12.1 records both numbers side by side, with the typical reasons for the gap.

Pack-on-demand throughput — OEM nameplate versus realistic field sustained.

Packsize X7 [89]	up to 1 200	≈ 720 to 960	Box-mix size variation, infeed gating, cycle losses on small boxes
CMC CartonWrap 1000evo [90]	up to 1 028	≈ 620 to 820	Print integration cycle, label-side timing, infeed pacing
Sparck CVP Impack [96]	up to 1 100	≈ 660 to 880	Cut-to-fit calculation, fanfold-perforation avoidance, varying box footprint
Kolbus Autobox AB 400 [97]	up to 1 200	≈ 720 to 960	Tape-closure timing, format changeover, manual reload of fanfold stacks

The challenge in pack-on-demand is twofold.

First, the creasing must be reproducibly precise across variable box sizes, that is, the variable-format problem.

Second, mechanical creasing acts immediately on a material that has had no recovery time after scoring; the box is folded seconds after the crease is impressed. Spring-back and fold fracture are particularly acute as a result, and the conditioning argument in Section 12.5 acquires its sharpest form in this configuration.

The fanfold problem: competing fold lines

A third challenge, specific to pack-on-demand, comes from the fanfold stock itself. Fanfold corrugated board is supplied in continuous, leporello-folded stacks, meaning the web is perforated or transversely creased at regular intervals by the converter. Without these intervals, the material could not be folded into a stack. At the packing machine, however, those pre-existing transverse lines become a source of defects. When the inline creasing tool impresses a new score at a position that does not coincide with a fanfold perforation, the blank carries two competing lines of reduced bending stiffness: the inline-created score and the factory-applied fanfold perforation.

During folding, the panel hinges along the line of least resistance. If the fanfold perforation lies within the panel or close to the intended score, the fold can migrate onto the perforation rather than following the inline score. The result is a skewed, displaced or double-folded panel; in severe cases, the panel breaks along the perforation, and the pack geometry is lost. The problem is most pronounced when box dimensions are chosen so that a fanfold transverse line falls within a few millimetres of an inline crease, or within a flat panel intended to remain unbent. Mitigation requires coordinated process planning. The cut-to-size algorithm must avoid placing dimensions that correspond with the fanfold pitch and the inline score geometry. The inline creasing must produce a score that is markedly weaker than the residual stiffness of the perforated line. Crease-line conditioning by waterscoring — discussed below — is particularly effective here: a well-conditioned inline score is reliably softer than a dry perforation and reasserts itself as the preferred hinge

The fundamental difference

Comparison of machine concepts — wrap-around and tray former versus pack-on-demand.

Creasing point	**In advance at the FFG or die-cutter**	**Inline, immediately before folding**
Blank feed	Flat, pre-scored blanks from a magazine	Continuous fanfold from reel or leporello
Fold geometry	Fixed per blank format; creases in both directions	Variable per box size; crease position variable
Adhesive	Hot melt (e.g. EVA), 4 to 5 g per side, sometimes interlocking	Hot melt or tape; sometimes adhesive-free (interlocking)
Speed	15 to 100 blanks per minute (beverage lines: 80–120 kcans/h)	OEM max 1 028 to 1 200 boxes/h; realistic sustained 600 to 960 (Table 12.1)
Spring-back regime	Moderate to high; depends on crease quality and storage time	High; no recovery time after creasing

The fundamental difference has direct consequences for crease-line conditioning. With pre-scored blanks, the task is to condition already-impressed crease lines: moisten before folding to reduce spring-back. With pack-on-demand, the task is process-integrated conditioning: moistening as part of the creasing operation itself, to improve crease quality from the outset and to ensure that the inline score, not a residual fanfold perforation, governs the fold.

12.3 Blank Types and Crease Directions

The European packaging industry uses a manageable number of standardised blank types, systematised in the FEFCO code [85]. To understand the creasing requirements, it is useful to know the geometry of these blanks and the direction of their crease lines.

The principal FEFCO types

FEFCO 0406 — Wrap-around. The classic format for beverage trays. A one-piece blank with integrated base and lid, whose side tabs overlap fully and are bonded with hot melt. Used for 6-, 12- and 24-pack multipacks of cans, bottles and jars. Crease lines run in both directions: longitudinal creases for side-panel folding and transverse creases for end-panel locking.

FEFCO 0421 — Self-locking tray. A tray with a self-locking end panel that requires no adhesive. Four side-panel creases plus corner solutions with interlocking tabs. Typical application: fruit and vegetable trays to the FEFCO Common Footprint standard [91] for optimised pallet loading.

FEFCO 0435 — Self-locking tray (variant). Comparable to 0421 but with a different corner solution. Frequently used for display trays in shelf-ready packaging (SRP), which serves simultaneously as transport packaging and retail presentation.

FEFCO 0700 series — Pre-glued variants. Blanks are bonded and flat-packed at the folder-gluer in the corrugated plant and delivered knocked down. Erected and filled at the case erector. Crease lines run exclusively in the longitudinal direction.

Crease-line direction and flute orientation

Crease lines on a typical tray blank run in both directions: longitudinal creases (parallel to the machine direction) and transverse creases (perpendicular to it). For a FEFCO 0406, this is typically four to six longitudinal creases for the side-panel pairs and two to four transverse creases for the end flaps. The orientation of the flutes relative to the crease has a decisive influence on folding behaviour. Creases that run transversely to the flute direction, across the flute tips, are less position-critical: the crease wheel meets similar conditions whether it lands on a tip or a valley, and spring-back is more moderate. Creases that run parallel to the flute direction are more critical: fold quality depends on whether the crease lands on a tip, in a valley or somewhere in between, and spring-back is significantly higher with greater scatter from blank to blank. The consequence for conditioning is direct: the flute-parallel creases benefit most from waterscoring, and the transverse creases also improve, particularly in multi-wall constructions.

12.4 Spring-Back: Why Crease Quality Determines Line Efficiency

Spring-back — the restoring force of a folded crease line — is the physical quantity that causes most problems at packing lines. It acts on every folding operation: when side panels are folded up, when end panels are closed, when corner locks are engaged.

The physics of spring-back

During creasing, the corrugated board is locally compressed: the flutes are crushed, the liners are plastically deformed, and the fibre-to-fibre bonds between plies are partially broken. Not all the deformation is plastic; however, an elastic component remains and generates the restoring force that acts to reopen the fold. Spring-back increases proportionally to the material's bending stiffness. Single-wall in kraftliner shows moderate spring-back; double-wall (BC, EB) in heavy liner grades generates restoring forces that push standard hot-melt bonds to their limits. With triple-wall, as used for heavy industrial packaging, conventional closure methods are frequently no longer sufficient.

Consequences at the packing line

Hot-melt consumption. The hot-melt must permanently overcome the spring-back force. At wrap-around packers, typical consumption is 4 to 5 g per bonded side, a figure largely set by spring-back. Reducing spring-back reduces the adhesive required.

Stoppages. Warped or inadequately creased blanks are the most common cause of stoppages at tray formers [84]. Board warp increases jam rate by 2 to 3 percentage points. Micro-stops, brief interruptions that are often not recorded, account for 15 to 20 per cent of total production-time loss. At 40 micro-stops of 3 minutes each per shift, 120 minutes are lost — a quarter of the available production time.

Reopens. When the spring-back force exceeds the hot-melt bond, the pack reopens after the bonding station. This causes jams at downstream stations or open packs on the pallet. The cause is insufficient creasing, too short a pressing time or too little adhesive — in each case mediated by spring-back.

OEE impact. Overall Equipment Effectiveness on typical packing lines is 60 to 75 per cent against a target of 75 to 85 per cent [83]. Crease-quality-related disruptions contribute significantly to the gap, but they are usually subsumed under "material defects" or "miscellaneous downtime" and remain invisible.

12.5 Waterscoring at the Packing Line: An Untapped Opportunity

Waterscoring is established at the folder-gluer. The Easyfold system [65] applies narrow fluid beads to the crease lines, the surfactant solution penetrates the liner fibres, bending stiffness drops locally and temporarily, folding becomes cleaner, spring-back falls and fold fracture is avoided. At packing lines, whether wrap-around packers, tray formers or pack-on-demand systems, the technology has not yet been adopted. The potential is considerable. Realising it requires a brief geometrical analysis.

Flute direction and machine direction

In typical wrap-around trays (FEFCO 0406), the flutes run parallel to the long side of the blank. The reason is stacking strength: in the finished tray, the flutes must stand vertically. Vertically fluted trays achieve approximately 20 per cent higher compression strength than horizontally fluted trays. This is the industry standard for European beverage and FMCG trays. The blank is generally fed into the packing machine with the flutes running perpendicular to the machine direction. Two consequences follow for the crease lines.

Longitudinal creases (parallel to machine direction). These creases run transversely across the flute tips. They typically serve the end-panel closures and lid flaps. Folding is less position-critical, and spring-back is moderate.

Transverse creases (perpendicular to machine direction). These creases run parallel to the flutes. They form the primary fold lines for the side panels — the load-bearing structure of the finished tray. Fold quality depends strongly on position relative to the flute tip or valley, spring-back is significantly higher, and scatter is greater.

This geometrical analysis is the basis for a two-stage implementation strategy.

Stage 1 — longitudinal application

Stage 1 limits the application to the machine direction. This is technically identical to the proven Easyfold solution at the folder-gluer. A nozzle bar above the transport path applies creasing fluid to the longitudinal crease lines as the blank passes beneath it. These longitudinal creases — the end-panel and closure creases — are not the most fold-critical, but they are functionally decisive: the end-panel closure determines whether the tray stays closed or reopens after the bonding station. Roughly half of all reopen events affect the end panels, because the adhesive load there is concentrated on a single hot-melt seam. Where conditioning reduces the spring-back of these creases, the failure rate at the end-panel closure falls immediately. Stage 1 reduces hot-melt consumption at the end panels, reduces the frequency of tray reopening after the bonding station, and improves closure quality (cleaner folds give more consistent dimensions and better stackability). Integration is straightforward: no new hardware concept, no transverse beam, no additional machine axis. Proven technology in a new application environment.

Stage 2 — transverse application: the full potential

Stage 2 adds a transverse nozzle bar that applies creasing fluid perpendicular to the machine direction onto the side-panel creases. These creases run parallel to the flutes and are the most fold-critical: spring-back is highest, the most fold fractures occur there, and the largest share of hot melt is needed to hold the side panels in position. The technical challenge of transverse application is synchronisation. The nozzle bar must place the fluid across the moving blank. Three options are technically feasible: traversing nozzles, a width-adjustable bar that follows the rule by mechanical coupling (the same approach already used to align stitching valves to the rule line, Section 11.4 Mechanism 1), or a brief stop phase during the transport cycle. All three add integration effort over Stage 1. Stage 2 unlocks the full potential: when both end-panel and side-panel creases are conditioned, the tray's total spring-back force decreases significantly. The combination of reduced hot-melt consumption, higher achievable speed and improved fold quality produces a multiplicative effect that justifies the investment in the extension.

Pack-on-demand: a special case

In pack-on-demand systems, creasing is carried out in-line. This creates the opportunity to apply creasing fluid immediately before or during the creasing operation: the fibres are conditioned before the mechanical load is applied. The effect is twofold — the crease itself becomes cleaner (less linear fracture, less fibre damage), and the subsequent folding benefits from the reduced spring-back force. Because pack-on-demand systems already have a multi-axis creasing station and box geometry varies from unit to unit, the application must be dynamically positionable, a requirement profile similar to the stitching operation at the folder-gluer (Section 10.4). The Stage logic does not apply here; the creasing fluid is applied to all crease lines because, in any case, the application is integrated into the multi-axis process.

Stages of crease-line conditioning at packing lines.

Creases treated	End panels and closures (transverse to flutes)	All creases including the side-panel creases parallel to flutes	All creases, applied during inline creasing
Hardware	One longitudinal nozzle bar (as in the folder-gluer	Longitudinal plus transverse nozzle bar, or an upstream conditioning station	Application integrated into the inline

	Easyfold installation)		creasing station
Integration effort	Low: proven technology in a new application environment	Medium: transverse synchronisation required	High: OEM partnership required for inline integration
Primary benefit	Fewer tray reopenings; reduced hot-melt at the end closure; better closure quality	Full spring-back reduction; maximum hot-melt saving; higher line speed	Better crease quality from the outset; less fold fracture; less waste
Investment	Low	Medium	Project-dependent (OEM)

12.6 The Economic Case for Crease-Line Conditioning

The economic case rests on four levers, summarised in Table 12.4.

Economic levers of crease-line conditioning at the packing line.

Hot-melt saving	Reduced spring-back → less adhesive needed to hold the fold permanently; with dotted application as the deposition method.	20 to 50 % reduction in hot-melt per pack (by analogy: up to 60 % cold-glue saving at the folder-gluer)

Stoppage reduction	More consistent folding → fewer jams, fewer micro-stops, fewer reopenings.	5 to 15 % fewer crease-quality-related stoppages
Throughput increase	Lower folding forces → higher achievable line speed without compromising fold quality.	5 to 10 % speed increase at the same waste level
Quality improvement	Cleaner folds → better dimensional accuracy → less waste, fewer complaints.	Reduction in waste rate; tolerance compliance below ± 1 mm vs. ± 1.5 mm

Stoppage cost: it depends on which line you mean

Unplanned stoppages in FMCG packing lines are the headline number that drives the investment case, and they are also the number most easily overstated by quoting a single figure. The widely-cited PMMI 2023 industry analysis [83] reports approximately 25,000 EUR per hour as a representative cost for a stopped FMCG line. This number is realistic for high-throughput beverage and high-volume household-care lines, where the combination of unit volume, the retail penalty for non-delivery, and amortised line cost is high. It is not a universal number. The stoppage cost on a line varies by an order of magnitude depending on what the line packs, who buys the output and how the upstream chain backs up; a representative bandwidth across the FMCG sector is 3,000 to 40,000 EUR per hour. Three regimes are useful to keep separate.

High-throughput beverage and household-care (25,000 to 40,000 EUR/h). Lines packing carbonated soft drinks, beer, water, laundry detergent or cleaning products. Unit volumes are large, the line typically runs 24/7 against a tight retail-delivery contract, and a stopped line backs up the filler, the labeller and the depalletiser within minutes. The PMMI 25 000 EUR/h figure is anchored in this regime.

Mid-volume FMCG (10 000 to 20 000 EUR/h). Lines packing dry food, biscuits, dairy, personal care, and shelf-stable groceries. Volumes are lower, the line is often shared across SKUs, and the operational cost of a stoppage is dominated by changeover and recovery rather than by retail penalty. Crease-related stoppages here are still consequential, but the per-hour cost is lower.

Low-volume speciality and contract packing (3,000 to 8 000 EUR/h). Co-packers, regional FMCG, cosmetics, premium frozen, slow-mover SKUs. Stoppages cost real money, but at a per-hour rate where the conditioning investment must justify itself on multi-year amortisation rather than on payback within months.

The implication for waterscoring at the packing line is that the strongest investment case sits at the high-throughput end — the regime in which the PMMI figure was originally measured. For mid-volume FMCG, the case is still positive, but the payback period lengthens; for low-volume speciality work, the conditioning argument competes with other line-improvement projects for a smaller absolute reward. The same physical intervention is the right answer in all three regimes; only its relative priority differs.

Material savings

The material side adds independently of the stoppage cost. At a hot-melt consumption of 4 to 5 g per bonded side and hot-melt prices of 2 to 4 EUR per kilogram, the absolute saving per pack is small — in the tenths-of-a-cent range. At beverage-tray volumes of millions of units per year, the small per-pack saving accumulates to a significant annual cost reduction; on lower-volume lines, the absolute number is correspondingly smaller. The waste-rate reduction (cleaner folds, fewer reopens reaching the case erector, fewer tolerance-out-of-band rejects) compounds the material number, and is the lever that scales most directly with line throughput across all three regimes.

12.7 Adjacent Technologies and Print-Line Integration

PSA tape closure

Pressure-sensitive adhesive (PSA) tape is a packing-line closure method that competes with hot-melt bonding in some applications, particularly hand-close use cases and low-throughput lines that lack hot-melt infrastructure. The tape is dispensed onto the closure seam, pressed in place, and the bond holds against the shipping load through the PSA's contact-bond strength. PSA tape closure is real and widely used — the Kolbus Autobox AB 400 [97], for instance, offers it as the standard closure mode for boxes too small or too irregular for cleanly metered hot-melt application. It is, however, not a substitute for crease-line conditioning for two reasons.

First, PSA closure adds a process step (tape dispensing) rather than reducing one; the spring-back force at the crease remains and is now resisted by the tape rather than by the hot-melt bond.

Second, PSA throughput rarely reaches the speeds at which hot-melt-closed beverage and household-care lines run; it is a hand-close-and-light-line method, not a competitor for the high-throughput regime where the economic case for waterscoring is strongest. Where PSA closure is used, the conditioning argument still applies on the crease side, but the closure-side savings shift from hot-melt consumption to tape consumption per pack.

Print-line integration

When inline waterscoring is added to a packing line that prints on the blank, either at the converter just before delivery, or inline within a pack-on-demand system that prints address labels or shipping marks on the outer surface, a time-window question arises. The fluid is applied to a band that is two to four millimetres wide, but the surfactant in the fluid lowers surface tension precisely so that the fluid spreads quickly into the substrate.

Any uncured ink in the application zone is at risk of migration, smearing, or local intensity loss. The mitigation is sequencing. Inline waterscoring should be applied downstream of the ink-fixation station: after UV curing for UV-cured inks, after the drying tunnel for water-based or solvent inks, after the hot-air bridge for hot-melt-pinned inks.

A pack-on-demand system in which the on-pack address printer sits between the creasing station and the closure has to be configured so that the conditioning fluid reaches the crease before the printer reaches the corresponding panel face, or so that the printer is upstream of the conditioning step on a different transport axis. In commercial installations, the issue is solved by layout rather than by chemistry: the fluid and the ink occupy different stations on the line, and the cure step between them is the buffer that prevents interaction. On pre-printed blanks delivered from the converter, the print is fully cured by the time the blank reaches the packing line, and the question does not arise.

Operationally, the applicator on a packing line is almost always mounted below the conveyor and sprays upward onto the underside of the blank, which is also the inside face once the blank is folded , the from-below geometry that Section 11.6 develops physically and that the surfactant chemistry permits. The print head, by contrast, sits above the conveyor and addresses its outer surface. Inside conditioning and outside printing, therefore, occupy different transport sides, and the timing question reduces to ensuring that whichever face is being treated has finished its previous process step before the next one engages.

A second print-line consideration concerns coated white liners on which the print sits over a clay or polymer coating. If the coating reduces the substrate's effective $Cobb_{60}$ below the lower bound of the standard waterscoring dosage range (Section 11.6), the fluid does not penetrate, and the conditioning effect is lost. The fix is to dose by Cobb of the actual surface, not by board grade alone, and to verify on a print-side specimen rather than on an unprinted offcut. Where the coating prevents penetration entirely, an alternative score-side application strategy , conditioning the inside of the blank rather than the printed outside, restores the effect, because the inside liner is rarely coated for print purposes and normally accepts the fluid.

The inside-application default also carries a print-side benefit at the packing line: where the outer liner is digitally printed with a continuous-design layout that crosses every fold (Section 11.4, Mechanism 3), the elasticity that waterscoring delivers eliminates the fine white hairline crack that otherwise runs along the printed fold edge — a defect that is invisible at the converter and visible only after folding, and that is one of the most common reasons premium-print pack-on-demand jobs are rejected at the print buyer's incoming-goods inspection.

12.8 Summary

Packing-line crease performance is determined by four pillars: the crease quality delivered by the corrugated plant, the spring-back behaviour the crease produces during folding, the hot-melt or tape closure that has to overcome that spring-back, and the line stoppages that crease-quality variability ultimately causes. The pre-scored pack-on-demand divide is real, but it is a divide of where the crease is impressed, not of what determines whether it folds well: in both cases, insufficient conditioning of the crease line is the upstream cause of the downstream defects.

Flute and machine direction set the conditioning strategy. In the typical wrap-around tray (FEFCO 0406), the flutes run perpendicular to the machine direction, so the longitudinal creases (end panels, transverse to the flutes) can be conditioned using the proven longitudinal Easyfold approach at a low integration cost (Stage 1). The transverse creases (side panels, parallel to the flutes) require a complementary transverse application (Stage 2) and achieve a larger share of the spring-back savings. Pack-on-demand, where the creasing is inline anyway, allows the conditioning fluid to be applied during the creasing operation itself, and the stage logic does not apply.

The economic case is dominated at the high-throughput end by stoppage cost, where the PMMI 2023 figure of approximately 25,000 EUR per hour [83] anchors the upper bracket of a sector bandwidth of 3,000 to 40,000 EUR per hour; the case still works at lower throughput, but the payback period lengthens. Hot-melt savings, throughput gains and waste minimisation add independently and compound across the line. Two adjacent considerations deserve a place in the planning: PSA tape closure, a real closure alternative for hand-close and low-speed regimes but not a substitute for crease conditioning; and print-line integration, a sequencing question (inline waterscoring downstream of ink fixation) rather than a chemistry question. The intervention is the same in every configuration; only its placement on the line differs.

Part V - Synthesis and Outlook

Chapter 13:
The Complete Picture: Interactions and the Integrated Concept

Introduction card

The final chapter brings diagnosis and treatment together. It explains how stitching and waterscoring interact, where their effects multiply, where their limits remain, and how plants can stage implementation across the corrugator, folder-gluer, packing line, palletising, and quality systems. It is the synthesis chapter for readers who need the technical argument, the business case, and the PPWR response in one place.

"How do stitching, waterscoring, adhesive monitoring, substrate control, and packing-line conditioning combine into one integrated answer to variation, downgauging, and PPWR?"

Economic impact: The business case is strongest where spring-back, narrow tabs, recycled grades, or lightweight boards create recurring stoppages and complaints. Stitching reduces adhesive mass and makes the glue line more damage-tolerant. Waterscoring reduces the load acting on that glue line. At the packing line, the case is dominated by stoppage cost, with high-volume FMCG lines carrying the strongest payback and lower-throughput lines requiring a more selective calculation.

Operational efficiency: The two main interventions act at different points. Stitching prevents local defects from propagating through the joint. Waterscoring reduces the spring-back force that creates the load in the first place. Together, they decouple variation from failure more effectively than either measure alone.

Quality: The integrated concept links the whole diagnostic chain. Corrugator-side defects must still be controlled upstream. Adhesive application must be monitored. Crease quality must be stabilised. Pallet overhang, interlocking patterns, and poor pallet quality must be avoided. The chapter therefore turns the book's mechanisms into a staged operating model.

Ecological and regulatory impact: PPWR makes material reduction unavoidable, but lighter material narrows the process window. The chapter argues that material minimisation and process precision must be treated together. Downgauging without process control increases failure risk. Downgauging with waterscoring, stitching, monitoring, and proper logistics control can reduce material use without sacrificing the protective function of the pack.

Limitations: Neither stitching nor waterscoring can rescue excessive washboarding, severe crushing, poor palletising, or unsuitable barrier-coated constructions. Waterscoring should usually be evaluated first where the substrate is the constraint. Stitching is the natural next step where the glue line must become more damage-tolerant.

Read in full: plant managers, operations directors, quality managers, packaging strategists, sustainability officers, and executives deciding on staged implementation.

Card sufficient: readers who have followed the earlier cards and need the final decision logic.

Prerequisite: Chapters 1 to 12.

Follow-up: implementation planning, plant-specific trials, and economic validation.

The preceding chapters examined defect formation at the corrugator (Chapter 2), the compounding effect at the folder-gluer (Chapter 3), the economic and ecological consequences (Chapters 4 and 5), the root-cause analysis (Chapter 6), the diagnosis of the secondary fracture (Chapter 8) and the chemistry of adhesive application (Chapter 9), then the two interventions — stitching (Chapter 10) and waterscoring (Chapters 11 and 12). Each covered a facet; none addressed the whole. This chapter brings the threads together. It shows where the individual measures interact, where synergies arise and where the limits remain. It translates the analysis into process-stage recommendations and gives an honest assessment of what is achievable today.

13.1 The Synergy: Stitching and Waterscoring

The two interventions in Part IV operate at different levels and complement each other. Stitching fragments the crack path and shifts the failure mode at the manufacturer's joint from propagation-driven to initiation-driven. Waterscoring reduces the spring-back load the joint must withstand in the seconds and minutes after folding, when the PVAc has reached only a fraction of its final strength. To use the medical analogy that has been earned by the diagnostic chapters: stitching is the treatment that contains the infection (crack propagation) once it starts; waterscoring is the intervention that reduces the load on the immune system in the first place. A clinician who only treats the symptom misses the cause; a clinician who only strengthens the constitution misses the moment of infection. Used together, the two address the same disease at two different points in its course.

The diagnosis-therapy loop is tighter than this metaphor suggests. Section 8.5 identified low elongation at break as the material-side signature of the secondary-fracture defect; the IGF 21804 BR sensitivity analysis [103] confirmed elongation as the only material parameter consistently linked to stable creasing across eleven corrugated grades. Section 11.4, Mechanism 3, identified increased liner elasticity as the active mechanism of waterscoring: the surfactant solution swells the cellulose fibres transversely by 20 to 30 per cent, hydrogen bonds are temporarily replaced by water, and the liner transitions from elastic-brittle to ductile in milliseconds. The diagnostic property and the therapeutic mechanism are the same variable, addressed by one intervention. The clinician's metaphor closes neatly at the level of measurement — the chapter that defines the disease names the cure.

Synergy 1: Additive safety reserve

Waterscoring reduces the spring-back force at the crease by a factor of three to four (Section 11.4, Mechanism 2; Auernhammer [104], Walker and Seffen [105], Navaranjan [106], Mrówczyński [82]). Stitching converts the strength distribution of the resulting joint from extreme-value to normal (Section 10.3 and Footnote 1 of Chapter 10). The two effects multiply: less load applied to a more redundant geometry. The empirical anchor is the PTS test report 35.756 (Chapter 10, Appendix A): three beads with 30 per cent stitching saving deliver +75 per cent in mean Factory Edge Strength against the three-bead continuous reference, with the coefficient of variation falling from 0.37 to 0.07. That number is itself measured against a dry-creased substrate; with waterscoring, the joint sees a smaller load to begin with, and the safety reserve compounds accordingly.

Synergy 2: Moisture dissipation through stitching gaps

A frequently raised concern about waterscoring is the presence of residual moisture at the crease line, which could migrate into the bonding zone and interfere with PVAc film formation. Stitching mitigates this risk by construction. The gaps between adhesive segments act as ventilation channels; residual moisture migrating from the crease into the MJ area does not encounter a continuous adhesive film but open zones through which it can evaporate. The intermittent application creates degrees of freedom for moisture transport that a continuous bead does not offer.

Synergy 3: Compensation for narrow tabs

PPWR-driven tab narrowing (Chapter 9) increases peel stress at the joint while reducing the bonded area. Waterscoring lowers the load by reducing springback; stitching raises the joint's tolerance against a single segment failing. Together, they let tabs that are unreliable when dry-creased work correctly, even at the lower end of the calibre range that PPWR pushes the industry toward.

Synergy 4: Stitching and dotting open up the stiffer grades

The fourth synergy is mechanical and follows directly from the load-and-redundancy structure of the previous three, but it deserves its own paragraph because it changes which grades are in the design space at all. The load that an MJ glue line must carry in the critical seconds and minutes after folding is dominated by spring-back, not by the eventual service load of the filled pack. In a conventional dry-scored configuration, the adhesive compensates for high spring-back by combining a sufficient green-tack per unit area with a generous bonded area. A continuous bead along a wide tab is, in this regime, not a design choice; it is an obligation imposed by the load.

Waterscoring changes the premise. When spring-back drops by a factor of three to four, the peel and cleavage stresses at the green bond drop by the same ratio, into a range that intermittent application can also hold. Stitching becomes viable on board grades where it was previously considered too risky, such as stiff double-wall constructions, high-performance liners, boards with elevated recycled content and correspondingly higher residual elasticity. The same logic extends to hot-melt dotting at the packing line (Chapter 10, callout in Section 10.3): the discrete-bond crack-arrest argument is the same physics in a different temperature regime, and the lower spring-back load is what makes the dotted pattern work where a continuous EVA bead used to be the only acceptable option.

The cost consequence follows directly. PVAc dispersion is roughly an order of magnitude more expensive per kilogram than waterscoring fluid. The continuous bead is an adhesive-intensive workaround for a mechanical problem; waterscoring addresses the mechanical problem at a tenth of the per-kilogram cost and lets the adhesive itself shift to an intermittent pattern that uses substantially less mass per pack. Stitching at 50 per cent duty cycle uses, to first order, half the adhesive of a continuous bead; dotting uses considerably less again. The combined saving , cheaper process aid plus reduced adhesive mass, is structural rather than marginal. On stiffer grades in particular, this synergy is what tips the business case: stitching and dotting cease to be cautious options reserved for cooperative materials and become the default, because the upstream waterscore has brought the material into cooperation.

Interaction matrix of stitching and waterscoring.

MJ strength under peel	Damage-tolerant; spring-back force unchanged	Spring-back reduced; joint conventional	Double reserve: less load against a more robust joint
Adhesive consumption	Up to 50 % reduction	Unchanged	50 % reduction with simultaneously improved performance
Crease quality and fold geometry	Not affected	Homogenised across flute positions; less scatter	Homogenised crease meets a damage-tolerant glue line
Liner fracture at crease	Not affected	Drastically reduced; print-side cracks eliminated	Reduced; crack-arrest pattern prevents propagation
Narrow tabs (PPWR)	Partial compensation through robustness	Partial compensation through reduced spring-back	Substantial compensation for the loss of bonding area
Residual moisture risk	Stitching gaps act as ventilation	Moisture is introduced locally and transiently	Moisture is introduced; gaps provide ventilation

13.2 The Limits of the Interventions

No intervention cures everything, and stitching combined with waterscoring is no exception. Five limits should be explicitly stated so that expectations remain realistic and investment decisions rest on complete information.

Limit 1: Corrugator-side defects remain

Neither stitching nor waterscoring can eliminate washboarding or warp generated at the corrugator. Waterscoring mitigates the consequences (better creasing despite washboarding, reduced spring-back despite warp), but the causes are upstream. A corrugator producing sustained washboarding with an amplitude above 100 µm or a warp beyond the symmetry tolerance has a problem that cannot be solved solely at the folder-gluer.

Limit 2: Substrate dependency remains for waterscoring

The waterscoring effect depends on the substrate's absorbance. $Cobb_{60}$ governs the dosage matrix (Section 11.6, Table 11.1): Kraftliner needs a higher dose and longer contact time, recycled liner accepts the fluid quickly but with a higher strike-through risk. The dose has to be matched per liner. Stitching, by contrast, works on all common liner types with a substrate-independent adhesive reduction; the synergy with waterscoring carries any liner-specific dose adjustment of the latter through to the system level.

Limit 3: Barrier-coated grades are excluded

Barrier-coated corrugated and carton — grades with a polymer or clay barrier on at least one face for grease, moisture, oxygen, or vapour resistance — sit entirely outside the dosage matrix. The barrier coating's defining property is water repellency; that is what makes the coating useful and also prevents the surfactant solution from penetrating. Where only the outer face is coated, inside application restores the effect (Sections 11.6 and 12.7); where both faces are coated, waterscoring is not applicable, and the crease must be solved by mechanical means alone (deeper score, narrower bead, multi-rule construction). Plant managers running mixed liner streams should treat barrier-coated lines as a separate process class.

Limit 4: Machine configuration

Waterscoring requires application units at the folder-gluer with nozzles specifically designed for the surfactant solution, positioned precisely on the crease lines. When mounted on the adjustable folding bars (Section 11.4, Mechanism 1), the setup reduces to a one-time mechanical alignment: the nozzles travel with the bars and automatically follow the format width, and the control system needs only the application length per format. Stitching requires an adhesive control system capable of intermittent application, which is standard in current-generation non-contact valve systems (Chapter 9) but not retrofittable to legacy wheel applicators or low-frequency valves (Chapter 10, Section 10.4 and Table 10.2). Both interventions assume the machine is valve-controlled at frequencies high enough to print the segmentation pattern cleanly at production speed.

Limit 5: Multidimensional interactions persist

The system remains multidimensional. Stitching and waterscoring improve two of its dimensions ,crack arrest at the joint and spring-back at the crease , but the others stay untouched. Storage humidity continues to act on BCT and warp. Palletising conditions (overhang, deck-board gaps) impose loads on the box independently of glue-line quality. Crushing at the corrugator destroys shear stiffness that no folder-gluer intervention can restore. The integrated concept shifts the system's working window; it does not rewrite the physics outside it.

13.3 Process-Stage Recommendations

The recommendations that follow summarise the findings of the preceding chapters into a practice-oriented guide. They are structured by process stage and are not a checklist in which every item must be implemented for the process to function. Most lines run, most of the time, without any of these interventions; what they address is the margin between a process that tolerates variation in substrate, climate and machine state, and one that does not.

Among the measures that follow, waterscoring occupies a distinct position. It is the only measure that directly extends the range of substrates that can be processed reliably — recycled liners with secondary-fibre content, testliners with high short-fibre fraction, stiff double- and triple-wall constructions, and coated white-tops on the inside-application route. Boards with elevated spring-back move into the centre of the process window once waterscoring is active. The other measures refine an already-working process; waterscoring enables a process on materials where the alternative is rejection of the grade. This distinction frames the recommendations below.

At the corrugator

Minimise adhesive add-on. Starch quantity at the corrugator is the strongest single driver of washboarding (Chapter 2). Every gram of excess starch per square metre amplifies the surface-wave amplitude almost linearly. The target is the minimum required to deliver a reliable fluting-to-liner bond, and nothing more. Modern dosing controls and inline moisture measurement permit a regulation level that earlier generations of corrugators could not reach.

Avoid crushing. The shear stiffness of the flute structure collapses disproportionately under crushing. The causes are excessive nip pressure at the drying section, transfer points, and the slitting station. A systematic review of nip pressures, ideally with pressure-measurement film, is the single most effective measure for preserving board stiffness.

Manage moisture symmetry. Symmetrical moisture content between the facings reduces warp. The target moisture differential between the outer and inner liner is below 1 per cent. Steam-shower systems upstream of the double facer correct asymmetric moisture profiles when detectable inline.

In storage

Climate control. Recycled liners respond disproportionately to climatic variation: at 90 per cent RH, the elastic modulus of recycled grades falls by approximately 70 per cent, against around 30 per cent for kraft liner [82, 106]. A controlled storage environment (20 to 25 °C, 45 to 65 per cent RH) pays off with less warp, more stable BCT, and more reliable MJ bonding.

Strict FIFO. Corrugated board ages, and moisture absorption and release during storage produce a hysteresis: a blank stored damp and then dried does not return to its original mechanical state. FIFO is not only logistical but also a quality measure.

At the folder-gluer

Waterscoring is the key intervention. Localised plasticisation of the crease line reduces local bending stiffness by a factor of three to four and spring-back in the same proportion (Section 11.4, Mechanism 2). The mechanical consequences are cumulative: fishtailing and gap deviation cease to be systematic defects (Mechanism 1, with the harmonisation argument in Section 11.5), peel load on the green MJ bond drops into the regime in which PVAc cohesion carries the joint comfortably, and the secondary fracture next to the crease (Section 8.5) does not initiate. The decisive point for material selection is that waterscoring extends the substrate window: grades that produce linear fracture, excessive spring-back or erratic fold geometry on dry scoring — high-recycled-content testliners, stiff heavy-wall boards, lightweight grades — become predictable once the score is conditioned. Dosage must be matched to liner type by $Cobb_{60}$ (Section 11.6); on coated outer liners, application is from the inside rather than the print face, and on barrier-coated grades it is not applied at all (Limit 3 above).

Wave-crest creasing for parallel scores. The IGF 21804 BR finite-element investigation [103] showed that for creases parallel to the flute direction, the wave-crest position (Bergrillung) imposes the lowest MD-stress on the upper liner during the rule's downward stroke; the valley position (Talrillung) imposes both a steeper initial rise and a higher peak; the intermediate position produces an inherently asymmetric crease bead. Parallel creases should therefore be performed as wave-crest creases by maintaining a constant longitudinal positioning of the corrugated board sheet relative to the rule. The valley case is mechanically the worst; the intermediate case is unrecoverable because the asymmetry is geometric.

Stitching is the natural next step. Once waterscoring has reduced the load, the joint can move from continuous bead to offset stitching with end-of-track adhesive guarantee (Section 10.4). The empirical case is the PTS 35.756 result quoted in Synergy 1: +75 per cent in Factory Edge Strength at -30 per cent adhesive, with a five-fold reduction in coefficient of variation. On stiffer grades, the same shift carries through to dotted hot-melt patterns at the packing line. The change is a software setting on a current-generation valve and integrated controller; on legacy folder-gluers, the upgrade path is in Table 10.2.

Adhesive application quality. High-closing-force valves with integrated water baths, purge firing, and multi-band IR sensors close the prevention-detection-response loop (Chapter 9). When the IR sensor coexists with inline waterscoring, the wavelength and spatial considerations of Section 9.6 must be incorporated into the integration plan. The sensor stream is also the input to SPC-based control of the application (Section 9.7), which converts threshold-alarm responses into trend-detection responses.

Adhesive matched to substrate. Standard PVAc dispersions are formulated for kraftliner. Substrates such as testliner, recycled liner and coated liner are better served by adapted formulations (Chapter 3); the supplier should know which substrates are running. Viscosity-temperature drift across the production day (Chapter 9, Table 9.2) is another variable to track and an input to the SPC layer.

At the packing line

Crease-line conditioning at the packing line is, in v18 terminology, the second deployment site for waterscoring — the under-recognised opportunity of Chapter 12. Two implementation stages apply to wrap-around and tray-former lines (Section 12.5): Stage 1 conditions the longitudinal end-panel creases with a single nozzle bar above the conveyor; Stage 2 adds a transverse bar for the side-panel creases. Pack-on-demand systems integrate conditioning into the inline creasing station itself.

Throughput honesty. Pack-on-demand throughput numbers in Section 12.2 (Table 12.1) distinguish OEM nameplate from realistic sustained: the headline 1,100 to 1,200 boxes/h are maxima under ideal conditions; field-realistic sustained values run 20 to 40 per cent below, depending on box mix, print integration, and operator response. Investment cases should be built on the lower bracket; the upper bracket is what the line does on its best day.

From-below applicator. Operationally, the conditioning fluid is applied below the conveyor onto the underside (inside) of the blank. The from-below geometry is permitted by the surfactant chemistry, which lowers the contact angle and draws the fluid into the fibre network within microseconds (Section 11.6 and the Washburn timescale of Section 11.3);

pure water cannot be applied from below for the same physical reason. The print head, where the line carries one, sits above the conveyor, so conditioning and print occupy different transport sides, and the timing question reduces to ink fixation upstream of the next process step (Section 12.7).

Print-side benefit. On digitally printed corrugated with a continuous-design layout that crosses every fold — the dominant aesthetic of premium-print pack-on-demand — the elasticity that waterscoring delivers eliminates the fine white hairline crack that otherwise appears along the printed fold edge after folding. This is the defect that print buyers see first; it is invisible at the converter and visible only after the pack is erected. The conditioning that prevents the inside-liner secondary fracture also prevents the print-side white-crack signature, by the same mechanism operating at both faces of the same fibre network (Section 11.4, Mechanism 3).

In palletising

Prevent overhang. An overhang of just 25 mm reduces BCT by 13 to 24 per cent; 50 mm reduces it by 29 to 40 per cent [43]. Pallet and box dimensions must be matched, a requirement that operations frequently overlook.

Columnar stacking, not interlocking. Interlocked stacking patterns reduce BCT by 35-40 per cent compared with columnar [44]. Where columnar stacking cannot guarantee pallet stability, stabilisation should come from stretch film or strapping, not from interlocking.

Pallet quality. Reusable pallets with splinters, protruding nails or uneven deck boards generate point loads that no box design accounts for. Systematic pallet inspection before loading reduces transit damage by a measurable amount.

In quality management

Cobb$_{60}$ as a process parameter. The Cobb value governs the interface between liner and adhesive, and the dosage window for waterscoring. It should be measured at every goods-in inspection and communicated to the folder-gluer. Deviations require dose adjustment for both adhesive and waterscoring.

Inline measurement. Modern folder-gluers offer inline monitoring — adhesive presence (multi-band IR), fold angle (laser scanner), and dimensional control (camera systems). The data enables early detection of drift before waste is generated. The IR data stream is also the input to the SPC layer in Section 9.7, which shifts the Level-3 response from reactive ejection to predictive trend control.

Defect-pattern database. Systematic logging of defect patterns — pop-opening, fishtailing, liner fracture, dimensional deviation, and secondary fracture — by the VDW–IGF code system of Section 8.2, along with their causes and remedies, is an underutilised tool. It connects the root-cause analysis of Chapter 6 with operational practice and accelerates the resolution of recurring problems. The Cross-Table of Section 8.10 is the diagnostic key for reading the database.

Reading the recommendations

A pragmatic priority follows from the structure above. Where the substrate is the constraint — high recycled content, where spring-back causes MJ failure or liner fracture, where narrow tabs under PPWR leave no tolerance — waterscoring should be the first measure evaluated. The other measures refine the process; they should not be mistaken for conditions without which the process cannot run.

13.4 Summary

Stitching and waterscoring are complementary, not alternative. Stitching makes the glue line more robust against defects; waterscoring reduces defects and load at the source, and extends the range of substrates that can be processed reliably. The combination produces disproportionate synergies, particularly with narrow tabs, lightweight boards and stiff or recycled grades. The PTS 35.756 measurement on B-flute (Chapter 10, Appendix A) gives the empirical anchor: +75 per cent in mean Factory Edge Strength at -30 per cent adhesive, with a five-fold reduction in coefficient of variation. The bond did not separate in any of the sixty test specimens — substrate failure throughout, which on B-flute means the joint is the stronger link of the assembly. On heavier flutes, the relative positions of the bond and the substrate may invert, and that experiment has not been published.

The recommended implementation path is parallel rather than sequential. Where the operation can take both changes, stitching and waterscoring should be introduced together: phasing them in separately captures a fraction of the benefit; phasing them in together captures the product of the two effects. Where only one is possible, the decision is unambiguous — waterscoring first. Waterscoring acts directly on the three metrics that drive the business case: it reduces waste, because the dominant defect modes tied to spring-back (fishtailing, gap deviation, liner fracture, pop-opening, secondary fracture) are attacked at their mechanical source; it reduces complaints, because the pack geometry and the print-side fold edge become reproducible; and it reduces cost, by widening the processable substrate window, by lowering the adhesive mass once spring-back no longer dictates a continuous bead, and by being itself roughly an order of magnitude cheaper per kilogram than PVAc. Stitching, important as it is, is a robustness measure; waterscoring is a direct lever on waste, complaints and cost.

The economic case at the packing line is dominated by stoppage cost (Chapter 12, Section 12.6). The PMMI 2023 figure [83] of approximately EUR 25,000 per hour anchors the upper end of an FMCG-sector bandwidth that ranges from EUR 3,000 to EUR 40,000 per hour, depending on whether the line packs high-volume beverage and household-care products, mid-volume food and personal care products, or low-volume speciality and contract products. The strongest investment case sits at the high-volume end, in the regime where the PMMI figure was originally measured; the case still works at lower throughput, but the payback period lengthens. The same physical intervention is the right answer in all three regimes; only its relative priority differs.

The limits of both interventions lie upstream and downstream. No folder-gluer intervention remedies excessive washboarding or crushing at the corrugator, and the best glue line is of little value if the box stands on a damaged pallet with 50 mm of overhang. Barrier-coated grades are excluded from waterscoring on the coated face, fully so where both faces are coated; the crease must be solved mechanically there. These boundaries must be respected. Within them, the combination of stitching and waterscoring — supported by current-generation non-contact application, multi-band IR monitoring with SPC-based response, and wave-crest creasing for parallel scores — defines the state of the art.

The PPWR paradox is solvable, but only through an integrated approach. Less material requires more precision; the two cannot be decoupled. Waterscoring and stitching, introduced in parallel, are that integrated approach. Where priorities force a choice, waterscoring is the measure that pays back first, fastest and most broadly — and the measure on which the rest of the design space rests.

Bibliography

[1] Beck, P.; Fischerauer, G. (2022): Modelling of Warp in Corrugated Cardboard by Homogenization and Inline Measurement. In: MDPI Applied Sciences. DOI: 10.3390/app12031684.

[2] Bronlund, J. et al. (2013): Modelling Steady-State Moisture Transport Through Corrugated Fibreboard Packaging. In: Packaging Technology and Science. DOI: 10.1002/pts.2025.

[3] Cornaggia, A. et al. (2023): Influence of Humidity and Temperature on Mechanical Properties of Corrugated Board. In: Materials. DOI: 10.15376/biores.18.4.7490-7509.

[4] Garbowski, T. et al. (2021): Crushing of Single-Walled Corrugated Board during Converting. In: Energies, 14(11), 3203. DOI: 10.3390/en14113203.

[5] Garbowski, T. (2024): The AI Revolution in Corrugated Board Production and Optimization. In: MDPI Machines.

[6] Garg, M. et al. (2016): Effect of Moisture Content on Printability of Corrugated Board. In: IPPTA Journal.

[7] Holmvall, M.; Uesaka, T. (2010): Print Quality vs. Process Conditions in Flexo Post-Print of Corrugated Board. In: Appita Journal.

[8] Jamsari, A. (2020): The Role of Flute Morphology in Mechanical Behaviour of Corrugated Fibreboard – A Numerical, Analytical and Empirical Study. PhD thesis, Massey University, Palmerston North, New Zealand.

[9] Johnson, M.; Popil, R. (2015): Corrugated Board Bonding Defect Visualization and Characterization. In: International Journal of Adhesion and Adhesives. DOI: 10.1016/j.ijadhadh.2014.12.030.

[10] Modzelewska, I. (2006): Climatic Conditions vs. Hygrostability and Strength Properties of Corrugated Board. In: Fibres & Textiles in Eastern Europe.

[11] Netz, E. (1998): Washboarding and Print Quality of Corrugated Board. In: Packaging Technology and Science, 11(6), 269–281.

[12] Pescoller, H. / PERET CORRCHECK (2016): Washboarding-Messung in der Qualitätskontrolle. Seminarunterlagen.

[13] Popil, R.E.; Schaepe, M.K. et al. (2007): Adhesive Level Effect on Corrugated Board Strength – Experiment and FE Modeling. In: International Progress in Paper Physics Seminar, Georgia Institute of Technology.

[14] Rogalka, M. et al. (2024): In-Situ Classification of Deformed Corrugated Board Using CNN. In: Sensors. DOI: 10.3390/s24041051.

[15] Szewczyk, W.; Glowacki, K. (2014): Effect of Humidity on Paper and Corrugated Board Strength Parameters. In: Fibres & Textiles in Eastern Europe.

[16] Tuomainen, T. et al. (2023): MRI of Water Vapor Absorption in Simulated Corrugated Board. In: Cellulose. DOI: 10.1007/s10570-023-05483-3.

[17] Vonk, N. et al. (2020): Full-Field Strain Fields and Shrinkage of Single Cellulose Fibers. In: Experimental Mechanics. DOI: 10.1515/npprj-2020-0071.

[18] Wade, M. et al. (2023): Does Flute Angle Influence Box Performance? In: BioResources. DOI: 10.1007/s10853-023-08941-2.

[19] Wendler, S. (2006): Washboarding of Corrugated Cardboard. Dissertation, RMIT University Melbourne.

[20] Cannella, F.; Dai, J. S. (2006): Crease Stiffness and Panel Compliance of Carton Folds. In: IMechE Journal of Mechanical Engineering Science, 220(12), 1833–1842. DOI: 10.1243/0954406JMES502.

[21] Conde, M. et al. (2012): Adhesive Joints of Corrugated Board Under Shear Loading – FEM Analysis. DOI: 10.1016/j.ijadhadh.2012.05.003.

[22] DIN 19304:2021-04: Prüfung von Wellpappe – Rillungen. Deutsches Institut für Normung.

[23] Ek, M. et al. (2009): Paper Physics – Handbook. De Gruyter. DOI: 10.1515/9783110213461.

[24] EP 0 381 080 A2 (1990): Pasting Device for a Glue Machine. Mitsubishi Heavy Industries. Geschwindigkeitssynchronisation der Auftragswalze.

[25] EP 2 279 856 B1 (2011): Leimverfahren und -vorrichtung für Wellpappebahnen. Reaktionskraftsensor-Regelung.

[26] EU Regulation 2025/40 (PPWR): Packaging and Packaging Waste Regulation. In force since 11 February 2025.

[27] Hubbe, M. A. (2024): Size Press Practices and Formulations Affecting Paper Properties and Process Efficiency: A Review. In: BioResources, 19(1), 1925–2002. DOI: 10.15376/biores.19.1.Hubbe.

[28] Mrówczyński, D. et al. (2023): Bending Stiffness of Unsymmetrical Multilayered Corrugated Board. In: BioResources, 18(4), 7611–7628. DOI: 10.15376/biores.18.4.7611-7628.

[29] Nagasawa, S.; Komiyama, Y. (2012): Effect of Scoring Condition on Creasing Characteristics of Double-Wall Corrugated Board. In: JSME, 6(7), 1263–1275. DOI: 10.1299/jamdsm.6.1263.

[30] Nica, W.; Chalmet, R.; Roman, J.; Juwet, M. (2020): Inline Corrugated Board Creasing. In: Procedia Manufacturing, 51, 501–504. DOI: 10.1016/j.promfg.2020.10.070.

[31] Pizzi, A.; Mittal, K. L. (2017): Handbook of Adhesive Technology, Third Edition. CRC Press, Boca Raton.

[32] Schneberger, G. L. (1983): Adhesives in Manufacturing. Marcel Dekker, New York.

[33] Thakkar, B. K. et al. (2009): Experimental and Numerical Investigation of Creasing in Corrugated Paperboard. DOI: 10.1080/14786430802342576.

[34] US 7 601 218 B2 (2009): Auto-Tracking Dispenser. Valco Cincinnati Inc. Dynamische Kartuschenanpassung an Werkstückplanarität.

[35] AICC: RSC Tolerances for Case Erectors and Packers. Technical Guidelines.

[36] Amcor (2020): How Much a Damaged Pack Can Really Cost Your Business.

[37] Amazon SIPP (2026): Ships in Product Packaging. Programme guidelines, valid for EU marketplaces (DE, FR, IT, ES).

[38] Baker, M. et al. (2016): Effect of Pallet Deckboard Stiffness and Unit Load Factors on Corrugated Box Compression Strength. Virginia Tech, Center for Unit Load Design. In: Packaging Technology and Science.

[39] FEFCO (2024): European Database for Corrugated Board Life Cycle Studies.

[40] Frank, B. (2014): Corrugated Box Compression – A Literature Survey. In: Packaging Technology and Science, 27(2), 105–128. DOI: 10.1002/pts.2019.

[41] EHI Retail Institute (2022): Inventurdifferenzen und Warenverluste im europäischen Einzelhandel. Total losses approx. 4.1 billion EUR/year.

[42] ISTA 3K / 3N: Test protocols for European FMCG supply chains and FTL transport. Supplemented by DIN 55440 (compressive resistance) and EUMOS 40509 (load stability).

[43] Kim, S. et al. (2023): Predicting the Effect of Pallet Overhang on the Box Compression Strength. In: Packaging Technology and Science. DOI: 10.1002/pts.2768.

[44] Kmita-Fudalej, G. et al. (2025): Effect of Stacking Configuration on Corrugated Board Compression Performance. In: BioResources.

[45] McKee, R. C.; Gander, J. W.; Wachuta, J. R. (1963): Compression Strength Formula for Corrugated Boxes. In: Paperboard Packaging, 48(8), 149–159.

[46] Quesenberry, C. B. (2020): The Effect of Pallet Top Deck Stiffness on Compression Strength of Unit Loads. Dissertation, Virginia Tech.

[47] Returnalyze (2023): The High Cost of Damaged Products.

[48] Singh, S. P. et al. (2008): Measurement and Analysis of Package Drops in Express Parcel Delivery. In: Journal of Testing and Evaluation, 36(4). DOI: 10.1520/JTE101390.

[49] BMEL (2024): Nationale Strategie zur Reduzierung der Lebensmittelverschwendung. 10.8 Mio. Tonnen/Jahr.

[50] CEPI (2023): Key Statistics – European Pulp & Paper Industry. Recycling rate corrugated board: 90.5% (2020).

[51] DHL eCommerce (2025): Returns Trends. Return logistics and CO_2 emissions.

[52] Eurostat / FEFCO (2023): Packaging Waste Statistics. Paper and board packaging: 82.5% recycling rate in the EU.

[53] FAO (2011): Global Food Losses and Food Waste. Rome.

[54] FEFCO (2023): New Corrugated Industry CO_2 Footprint. 491 kg CO_2e/tonne (2021 data).

[55] FEFCO/CCB (2023): European Database for Corrugated Board Life Cycle Studies.

[56] McKinsey & Company (2021): True Packaging Sustainability – Understanding the Performance Trade-Offs.

[57] Packaging Digest (2024): Mitigating Packaging Damage in the Supply Chain. 11% damage rate at distribution centres.

[58] Sendcloud (2024): Stand der Retouren im E-Commerce Deutschland. Return rates and return costs.

[59] Thünen-Institut (2023): Lebensmittelverschwendung als Umweltbelastung. Approx. 8 Mio. t CO_2/year avoidable.

[60] Umweltbundesamt (2023): Lebensmittelabfälle verursachen vier Prozent der deutschen Treibhausgasemissionen.

[61] Williams, H. et al. (2012): Reasons for Household Food Waste with Special Attention to Packaging. In: Journal of Cleaner Production, 24, 141–148. DOI: 10.1016/j.jclepro.2011.11.044.

[62] ECHA: SVHC Candidate List under REACH Regulation (EC) No 1907/2006. Borax listed as Substance of Very High Concern.

[63] EP 3 617 399 A1: Borax-reduced starch adhesive for corrugated board production with MFC additive.

[64] Ramos-Gonzalez, R. (2015): Flute Tip Densification and Its Effects on Corrugated Board Properties. Dissertation.

[65] Baumer hhs GmbH (2025): CorrBox Solution evo – Product documentation. Krefeld.

[66] Baumer hhs GmbH (2024): PX 1000 – The Most Powerful Cold Glue Valve Available. Data sheet.

[67] Baumer hhs GmbH (2023): GDX 1000 Cold Glue Sensor – The New Standard in Adhesive Monitoring. Blog post.

[68] Intercol Adhesives (2020): Perfect Gluing – Troubleshooting Guide for Folder-Gluer Applications. Amsterdam.

[69] Combi Packaging Systems (2024): Most Troubles with Automated Box Forming Aren't the Machine – It's the Corrugated. Technical blog.

[70] Lanco Adhesives (2024): Troubleshooting Common Glue Lap Issues in Corrugated Sheet Plants. Technical bulletin.

[71] Schneberger, G. L.: Adhesives in Manufacturing – Chapter 16: Water-Based Adhesives.

[72] Pizzi, A.; Mittal, K. L. (2003): Handbook of Adhesive Technology. 2nd ed. Marcel Dekker, New York.

[73] CEPI Containerboard (2020): European List of Corrugated Base Papers. Brussels.

[74] Hallbäck, N. et al. (2014): Finite Element Analysis of Hot Melt Adhesive Joints in Carton Board. In: Packaging Technology and Science, 27(9), 701–712.

[75] ISO 535:2023: Paper and board – Determination of water absorptiveness – Cobb method.

[76] Griffith, A. A. (1921): The Phenomena of Rupture and Flow in Solids. In: Philosophical Transactions of the Royal Society, 221A, 163–198.

[77] Irwin, G. R. (1957): Analysis of Stresses and Strains Near the End of a Crack Traversing a Plate. In: Journal of Applied Mechanics, 24, 361–364.

[78] Paris, P. C.; Erdogan, F. (1963): A Critical Analysis of Crack Propagation Laws. In: Journal of Basic Engineering, 85(4), 528–534. DOI: 10.1115/1.3656900.

[79] Rahman, A. A. et al. (2006): Finite Element Analysis of the Deformation of Corrugated Board during the Scoring Process. In: Journal of Manufacturing Science and Engineering.

[80] Vonk, N. H. et al. (2021): Full-field Hygro-expansion Characterization of Single Softwood and Hardwood Pulp Fibers. In: Nordic Pulp & Paper Research Journal, 36(1), 61–74. DOI: 10.1515/npprj-2020-0071.

[81] Washburn, E. W. (1921): The Dynamics of Capillary Flow. In: Physical Review, 17(3), 273–283.

[82] Mrówczyński, D. et al. (2024): Impact of Temperature and Humidity on Key Mechanical Properties of Corrugated Board. In: Applied Sciences, 14(24), 12012. DOI: 10.3390/app142412012.

[83] PMMI (2023): State of the Industry – End-of-Line Packaging. Reston, VA.

[84] Lantech Inc. (2024): Why Do Case Erectors Jam? Technical White Paper.

[85] FEFCO (2022): International Fibreboard Case Code – 12th Edition. Brussels.

[86] AICC/PMMI (2023): RSC Tolerances for Case Erectors and Packers. Technical Guideline.

[87] TAPPI T 829: Scoring Quality and Bending Stiffness of Corrugated Board.

[88] Nagasawa, S. et al. (2013): Effect of Scoring Condition on Creasing Characteristics of Double-Wall Corrugated Board. In: JAMDSM, 7(1).

[89] Packsize International (2025): X7 Automated Packaging System – Product Documentation. Salt Lake City.

[90] CMC Machinery (2024): CartonWrap 1000evo – Technical Specifications. Città di Castello.

[91] FEFCO (2020): Common Footprint Standard. Brussels.

[92] Valco Melton (2024): Wrap-Around Cartoning – Adhesive Application Guide. Cincinnati.

[93] EUMOS 40509:2020: Test Method for Rigidity of Unit Loads.

[94] Schubert GmbH (2024): TLM Wrap-Around Packer – Technical Documentation. Crailsheim.

[95] FEFCO TS 702 (2019): Technical Specifications for Converting Equipment – Fishtailing & Gap Variation. Brussels.

[96] Sparck Technologies (2024): CVP Impack – Automated Fit-to-Size Packaging. Technical Documentation. Eersel.

[97] Kolbus Autobox GmbH (2024): Autobox AB 400 – Automated Box-on-Demand System. Technical Documentation. Rahden.

[98] Beex, L. A. A.; Peerlings, R. H. J. (2009): An Experimental and Computational Study of Laminated Paperboard Creasing and Folding. In: Philosophical Magazine, 89 (34 to 36), pp. 3413 to 3436. DOI: 10.1080/14786430903123450.

[99] C and T Matrix Ltd. (n. d.): A Guide to Creasing Matrix Selection and Creasing Practice for Corrugated Board. Technical booklet, C and T Matrix, Gainsborough, United Kingdom.

[100] Huang, H.; Hagman, A.; Nygårds, M. (2014): Quasi Static Analysis of Creasing and Folding for Three Paperboards. In: Mechanics of Materials, 69 (1), pp. 11 to 34. DOI: 10.1016/j.mechmat.2013.09.016.

[101] Nagasawa, S.; Fukuzawa, Y.; Yamaguchi, T.; Tsukatani, S.; Katayama, I. (2003): Effect of Crease Depth and Crease Deviation on Folding Deformation Characteristics of Coated Paperboard. In: Journal of Materials Processing Technology, 140 (1 to 3), pp. 157 to 162. DOI: 10.1016/S0924-0136(03)00825-2.

[102] Garbowski, T.; Gajewski, T.; Grabski, J. K. (2020): The Role of Buckling in the Estimation of Compressive Strength of Corrugated Cardboard Boxes. In: Materials, 13(20), 4578. DOI: 10.3390/ma13204578.

[103] Harling, A.; Schlimpert, F. (2024): Praxisorientierte Modellierung und Validierung der Zusammenhänge zwischen den Materialkennwerten von Wellpappe, den Parametern der Rillwerkzeuge und dem Rillergebnis (Rillen Wellpappe). IGF-Forschungsbericht 21804 BR. Papiertechnische Stiftung (PTS), Heidenau / Hochschule für Technik, Wirtschaft und Kultur Leipzig (HTWK).

[104] Auernhammer, J.; Keil, T.; Lin, B.; Schäfer, J.-L.; Xu, B.-X.; Biesalski, M.; Stark, R. W. (2021): Mapping Humidity-Dependent Mechanical Properties of a Single Cellulose Fibre. In: Cellulose, 28(13), 8313–8332. DOI: 10.1007/s10570-021-04058-4.

[105] Walker, M. G.; Seffen, K. A. (2019): The Flexural Mechanics of Creased Thin Strips. In: International Journal of Solids and Structures, 167, 192–201. DOI: 10.1016/j.ijsolstr.2019.03.015.

[106] Navaranjan, N.; Dickson, A.; Paltakari, J.; Ilmonen, K. (2013): Humidity Effect on Compressive Deformation and Failure of Recycled and Virgin Layered Corrugated Paperboard Structures. In: Composites: Part B, 45(1), 965–971. DOI: 10.1016/j.compositesb.2012.05.037.

[107] Robertsson, K.; Jacobsson, E.; Wallin, M.; Borgqvist, E.; Ristinmaa, M.; Tryding, J. (2023): A Continuum Damage Model for Creasing and Folding of Paperboard. Division of Solid Mechanics, Lund University, Sweden / Tetra Pak Packaging Solutions, Lund, Sweden.

[108] Csavajda, P.; Böröcz, P. (2025): Can It Still Be Used? Compression Strength Recovery of Corrugated Boxes After Moisture Exposure. In: Chemical Engineering Transactions, 121, 43–48. DOI: 10.3303/CET25121008.